Treating
LIFE-THREATENING
BLEEDINGS

TREATING
LIFE-THREATENING BLEEDINGS

Development of Recombinant Coagulation Factor VIIa

ULLA K.E. HEDNER MD PHD

Emeritus Professor of Hemostasis Research, University of Lund, Lund, Sweden

ACADEMIC PRESS

An imprint of Elsevier

Academic Press is an imprint of Elsevier
125 London Wall, London EC2Y 5AS, United Kingdom
525 B Street, Suite 1800, San Diego, CA 92101-4495, United States
50 Hampshire Street, 5th Floor, Cambridge, MA 02139, United States
The Boulevard, Langford Lane, Kidlington, Oxford OX5 1GB, United Kingdom

British Library Cataloguing-in-Publication Data
A catalogue record for this book is available from the British Library

Library of Congress Cataloging-in-Publication Data
A catalog record for this book is available from the Library of Congress

ISBN: 978-0-12-812439-0

For Information on all Academic Press publications
visit our website at https://www.elsevier.com/books-and-journals

Working together
to grow libraries in
developing countries

www.elsevier.com • www.bookaid.org

Publisher: Mica Haley
Acquisition Editor: Tari Broderick
Editorial Project Manager: Renata R. Rodrigues
Production Project Manager: Karen East and Kirsty Halterman
Cover Designer: Matthew Limbert

Typeset by MPS Limited, Chennai, India

CONTENTS

ABOUT THE AUTHOR

Ulla Hedner received her MD in 1966 and PhD in 1974 at Lund University, Sweden, and worked in the Comprehensive Hemophilia Care Center, Malmö in 1972–83. In 1978 to 1980, she served as a visiting scientist in the laboratory of Earl Davie, Professor of Biochemistry, at the University of Washington, Seattle, United States. She was recruited in 1983 by the Danish pharmaceutical company Novo Nordisk A/S

Photo: Peter Kroon

(Research & Development) and became Professor of Hemostasis Research in Gothenburg and Lund in 1988. During this time she was the mentor for five PhD students and the examiner of eight PhD students at different universities. She is a member of the American Society of Hematology, the International Society of Thrombosis Hemostasis (ISTH), and the World Fed of Haemophilia. Between 1996 and 2003 she was a member of the ISTH Council and in 2015 she was awarded one of the Distinguished Career Awards, a part of the "17th Biennial Awards for Contributions to Hemostasis" of the ISTH. She is now a member of the Novo Nordisk Hemophilia Foundation Council from 2006 and has authored 270 publications in peer-reviewed journals. Her family includes husband, and daughter and her family including two grandchildren.

PREFACE AND ACKNOWLEDGMENTS

Many have been at my side during the work with this book which describes the development of a hemostatic drug based on a new concept of hemostasis. It is the result of a close collaboration between the clinic and university and later also including pharmaceutical industry. All three, and not the least the collaboration between three of them, are absolutely necessary for a successful development of a new medicine.

Early during my years in Medical School, I became convinced about the importance of a near connection between basic science and the clinic. Later this was supported by my Professor of Internal Medicine at the University of Lund, Sweden, Jan Waldenström. He stressed the importance of carefully listening to a patient for correct diagnosis and therapy. In the patient, he always saw the biological problem, which he felt as a challenge to solve.

With this background and my own interest in biochemistry, it was clear to me that the obvious need for an improved therapy in hemophilia patients with inhibitors against the coagulation factors they were lacking had to be addressed in a tight collaboration between basic science and the clinical observations. Thus I became the bridge between these areas. My environment at the Department of Bleeding Disorders at the University Hospital of Malmö, Lund University, Sweden, with Professor Inga Marie Nilsson as the head was an excellent place for this.

Here I also met my first "key patient" who was successfully treated with purified coagulation factor FVIIa. Thus this patient became the "proof of principle" to me that the administration of pharmacological doses of pure FVIIa induced hemostasis in a hemophilia patient with inhibitors.

The other important factor that made this development feasible was my international research contacts, scientists I met at international meetings in the 1970s. Among those Harold Roberts, Professor of Medicine and Pathology, University of North Carolina of Chapel Hill School of Medicine, Chapel Hill, North Carolina, whose haemophilia clinic was one of the best worldwide. In addition, he had an excellent research group focusing on hemophilia research.

Dr. Roberts has during all my years in the hemostasis research been an important supporter, who has helped with interesting scientific discussions and a lot of encouragement during my many periods of despair and doubt about the whole endeavor at getting rFVIIa out to the patients I knew needed it. Over the years, Harold and his wife Mary became my very good personal friends which I am very thankful for.

Among other scientists who meant a lot to me, Earl Davie, Professor of Biochemistry at the University of Washington, Seattle, United States, has a special position. The possibility to spend the time in his laboratory with its special atmosphere of kindness and willingness to help and to discuss various issues was very important for me. I owe him a lot for his patience and encouragement over the years.

In Earl's laboratory, I also met Walter Kisiel who spent a year in our Hemophilia Clinic in 1980–81 after I was back from Seattle. He purified FVIIa and together we developed the first sample of absolute pure human FVIIa. Walter Kisiel later was a consultant to us at the start of the development of recombinant FVIIa.

I must also mention another mentor and great example from whom I learned a lot, Samuel I. Rapaport, Professor of Medicine and Pathology, UCSD School of Medicine, La Jolla, CA, United States. His publication in 1979 where he suggested that "substantial amounts of activated factor VII in the circulating blood could exert a hemostatic effect..." or at least this could not be excluded gave me a substantial hope that I may be right in my thoughts along the same lines and gave me energy to continue. Dr. Rapaport also became a good friend of mine over the years. After Dr. Rapaport's retirement, our hemostasis research group at Novo Nordisk had a close collaboration with one of his younger colleagues, L. Vijaya M. Rao and his wife, Usha Pendurhti.

The next important step in my carrier and in the development of rFVIIa was my encounter with the pharmaceutical industry, first in the shape of Novo Industri A/S and later as Novo Nordisk A/S, Denmark. My experience of pharmaceutical industry before that was very limited, but I was intrigued by the challenge to establish a research laboratory within hemostasis which they offered me when recruiting me in 1983. I stayed with them for the rest of my professional life. During these more than 25 years, I learned a lot, not the least about the conditions pharmaceutical industry is working under in the development of new medicine.

At this time, I was convinced that the facilities and knowledge of a pharmaceutical industry with experience in large scale fermentation were necessary, if rFVIIa ever should become available in the market place and get out to patients in need for it. I was impressed by the willingness of

Novo Industry to discuss the possibility to initiate a project with the aim to develop a recombinant drug for a comparatively small group of patients—the hemophilia patients with inhibitors.

Already very early, it was clear that such a project would require extensive investments in new technology and after previously having failed to raise any interest in two other companies, I was so surprised and happy to find a different attitude and more of interest in the challenge. I was especially impressed by the CEO, Mads Øvlisen, the Chief Officer of Science, Professor Ulrik Lassen, and the corresponding head of research of the enzyme part of Novo Industry, Knud Aunstrup, who all quickly got the idea of what might be achieved for these patients.

During the 10 years it took to get the rFVIIa approved in Europe my role was to keep the bridge between the research and the clinic. This was important in order to make sure that the focus on the patients was not lost. Also in this context, it was necessary to emphasize the importance of close collaboration between the basic research and the clinic. I met so many very qualified and dedicated people during these years of development involving so many different areas who all helped to make the rFVIIa a success. Many of the names of these people are mentioned in the book.

I also learned how to survive in spite of many extensive reorganizations of the company. The importance of a genuine and professional leader became obvious. I also have learned about the long-lasting work necessary after approval of a new drug to make sure the drug is spread among the patients in need of it. This requires education of the doctors in the area but also an increased awareness among society and payers. During the last ten years, I have worked with the Novo Nordisk affiliates all over the world on these issues and again I have learned a lot, met many knowledgeable, nice, and warm persons both among the Novo Nordisk affiliate people and among doctors.

Finally, I would like to thank my family who has encouraged me to write this book and has put up with all my traveling, long working hours, etc. My daughter and her family, and even my two grandchildren have kept track on the process of this book and advised me about how to make it attractive. In this context, I also want to give a special thanks to my daughter, Karin, who helped me to find the correct Talmud page describing the bleeding problems typical for hemophilia. She also checked that the Hebrew text was correctly translated. However, without the support and pushing from my husband and his critical readings of my text, I am afraid this book would never have been finalized. These last months he has taken care of all the technique without which nothing seems to be made feasible these days, and without him this would never have been achieved. He deserves many delicate dinners!

CHAPTER ONE

Classical Bleeding Disease (Hemophilia)

Contents

1.1 THE HISTORY OF HEMOPHILIA

Classical bleeding disease was first described in 1803 by the American physician John Conrad Otto (1774−1844) under the title "An account of an hemorrhagic disposition existing in certain families" [1]. By studying his patients and their families, he found the following characteristics that are still relevant:

1. Hereditary bleeding disease
2. Only men become ill, but the disposition is carried by women without symptoms
3. Prolonged coagulation time (time for the blood to clot)
4. Joint bleedings

Ten years after J.C. Otto's description, John Hay published an article in which he traced a family with bleeding disease 100 years back in time to the 1700s. He concluded that in this family, the disease never presented itself in the children of a bleeder, only in the male grandchildren [2].

However, there are earlier descriptions of bleeding disease, which most likely have been cases of hemophilia. The oldest known one is found in the Babylonian Talmud (Tractate Yebamoth 64b) where it can be read "for it was taught: If she circumcised her first child and he died, and a second one also died, she must not circumcise her third child. These are the words of Rabbi Judah the Patriarch, redactor of the

Treating Life-Threatening Bleedings.
DOI: http://dx.doi.org/10.1016/B978-0-12-812439-0.00001-0

Mishnah, the second century compilation of Jewish law." In the tractate Hullin 47b, Raba Nathan gives the same advice to mothers who have lost two sons through bleeding in conjunction with circumcision [3] (Fig. 1.1).

Similar recommendations are found in Jewish writings from the later middle ages [4]. It is interesting to note that already in the early years CE, rabbis were well acquainted with the heredity of this serious bleeding disease. The term hemophilia (meaning "love of blood") was first used in 1828 by F. Hopff in his doctoral dissertation [5].

In the 1800s and early 1900s, there were lively discussions about the cause of this bleeding disease. A. Schmidt (1892) described the normal coagulation process as an enzymatic process consisting of two phases: the first phase leading to the formation of thrombin in the presence of a tissue factor, also called as thrombokinase, thromboplastin, thrombozyme, or cytozyme, and calcium. The second phase involved the transformation of fibrinogen into fibrin in the presence of thrombin. This view of the coagulation mechanism was shared by the majority of researchers, although they had different names for the different components [6].

In the publication from 1893, Z. Manteuffel demonstrated that the prolonged clotting time in patients with hemophilia was normalized by the addition of the so-called thromboplastin (tissue extract) and suggested that hemophilia was caused by a lack of thromboplastin [7]. Platelets as a source of thromboplastin and prothrombin was also discussed [8]. Tissue liquid from various organs, especially the brain, was considered as an important source of thromboplastin by several researchers.

As early as 1911, T. Addis suggested that bleeding disease was caused by the failure of the patients' prothrombin to be transformed normally into thrombin [9]. However, such a defect could not be proved at that time. In the late 1930s, American and European researchers found that the addition of normal plasma shortened the clotting time of the blood from patients with hemophilia. To achieve a more marked effect, an attempt was made to treat normal plasma in different ways to produce plasma components containing more of the elements lacking in hemophilia. In 1936, K. Lenggenhagen suggested that the reason for the poor coagulation ability in hemophilia was the slow transformation of prothrombin into thrombin [10]. This was confirmed in the year of 1939 by K.M. Brinkhous [11].

Already in the early descriptions of hemophilia, it was suggested that it was a recessive hereditary condition transmitted by symptom-free

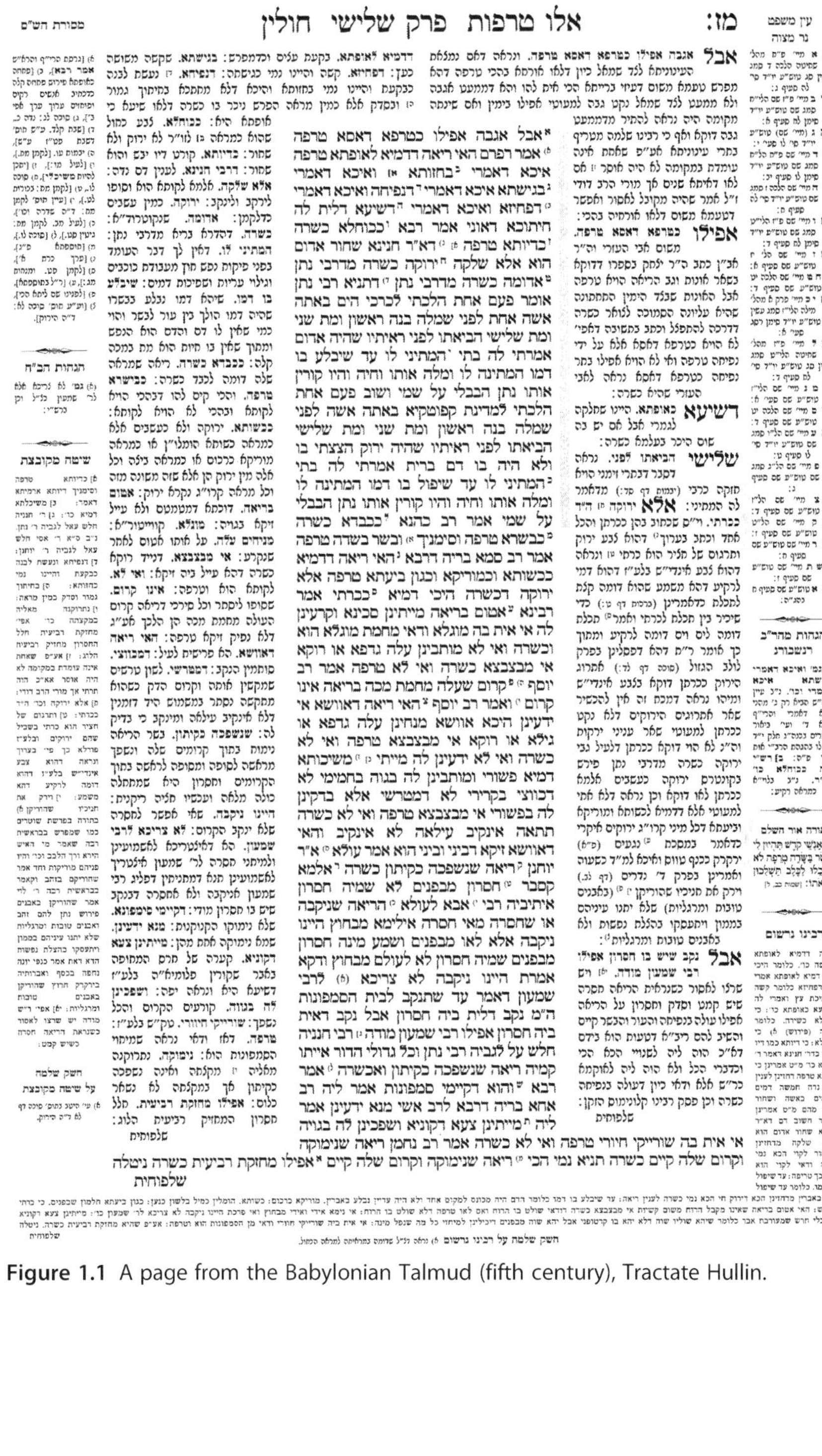
מז: אלו טרפות פרק שלישי חולין

אבל אגבה אפילו כטרפא דאסא טרפה אמר רפרם האי ריאה דדמיא לאופתא טרפה איכא דאמרי בחזותא ואיכא דאמרי בגישתא איכא דאמרי דנפיחה ואיכא דאמרי דפחיזא ואיכא דאמרי דשיעא דלית לה חיתוכא דאוני אמר רבא ככוחלא כשרה כדיותא טרפה דא"ר חנינא שחור אדום הוא אלא שלקה ירוקה כשרה מדרבי נתן אדומה כשרה מדרבי נתן דתניא רבי נתן אומר פעם אחת הלכתי לכרכי הים באתה אשה אחת לפני שמלה בנה ראשון ומת שני ומת שלישי הביאתו לפני ראיתיו שהיה אדום אמרתי לה בתי המתיני לו עד שיבלע בו דמו המתינה לו ומלה אותו וחיה והיו קורין אותו נתן הבבלי על שמי ושוב פעם אחת הלכתי למדינת קפוטקיא באתה אשה לפני שמלה בנה ראשון ומת שני ומת שלישי הביאתו לפני ראיתיו שהיה ירוק הצצתי בו ולא היה בו דם ברית אמרתי לה בתי המתיני לו עד שיפול בו דמו המתינה לו ומלה אותו וחיה והיו קורין אותו נתן הבבלי על שמי אמר רב כהנא כבבדא כשרה כבשרא טרפה וסימנך ובשר בשדה טרפה אמר רב סמא בריה דרבא האי ריאה דדמיא כבשותא וכמוריקא וכגון ביעתא טרפה אלא ירוקה דכשרה היכי דמיא ככרתי אמר רבינא אטום בריאה מייתינן סכינא וקרעינן לה אי אית בה מוגלא ודאי מחמת מוגלא הוא וכשרה ואי לא מותבינן עלה גדפא או רוקא אי מבצבצא כשרה ואי לא טרפה אמר רב יוסף קרום שעלה מחמת מכה בריאה אינו קרום ואמר רב יוסף האי ריאה דאוושא אי ידעינן היכא אוושא מנחינן עלה גדפא או גילא או רוקא אי מבצבצא טרפה ואי לא כשרה ואי לא ידעינן לה מייתי משיכותא דמיא פשורי ומותבינן לה בגוה בחמימי לא דכווצי בקרירי לא דמטרשי אלא בדקינן לה בפשורי אי מבצבצא טרפה ואי לא כשרה תתאה אינקיב עילאה לא אינקיב והאי דאוושא זיקא דביני וביני הוא אמר עולא א"ר יוחנן ריאה שנשפכה כקיתון כשרה אלמא קסבר חסרון מבפנים לא שמיה חסרון איתיביה רבי אבא לעולא הריאה שניקבה או שחסרה מאי חסרה אילימא מבחוץ היינו ניקבה אלא לאו מבפנים ושמע מינה חסרון מבפנים שמיה חסרון לא לעולם מבחוץ ודקא אמרת היינו ניקבה לא צריכא לרבי שמעון דאמר עד שתנקב לבית הסמפונות ה"מ נקב דלית ביה חסרון אבל נקב דאית ביה חסרון אפילו רבי שמעון מודה רבי חנניה חלש על לגביה רבי נתן וכל גדולי הדור אייתו קמיה ריאה שנשפכה כקיתון ואכשרה אמר רבא והוא דקיימי סמפונות אמר ליה רב אחא בריה דרבא לרב אשי מנא ידעינן אמר ליה מייתינן צעא דקוניא ושפכינן לה בגויה אי אית בה שורייקי חיורי טרפה ואי לא כשרה אמר רב נחמן ריאה שנימוקה וקרום שלה קיים כשרה תניא נמי הכי ריאה שנימוקה וקרום שלה קיים אפילו מחזקת רביעית כשרה ניטלה

Figure 1.1 A page from the Babylonian Talmud (fifth century), Tractate Hullin.

women and affecting only men. Queen Victoria of Great Britain (1819–1901) is probably the most well-known carrier of hemophilia. She gave birth to a son with hemophilia, Leopold (1853–1884), her fourth son and eighth child (Fig. 1.2). There has been much discussion as to whether Queen Victoria's gene for hemophilia had been inherited from her mother's side of the family or whether it was the result of a spontaneous mutation. In a biography, *Prince Leopold* by Charlotte Zeepvat, published in 1998, it was noted that several boys in Queen Victoria's family, dating back to around the year 1700, had died at an early age; several of them died as a result of "stroke," which is rare in young person. This is, of course, not the proof that they suffered from hemophilia, but it is an interesting observation [12].

At the age of 3 years, Leopold was described as being prone to more bruising than his siblings. It was also noted that he was disabled for several days or even weeks after minor accidents. He does not seem to have elicited much sympathy at home, and the queen was often impatient and complained about her youngest son, which is not an uncommon behavior in mothers who have a son with a serious disease. Although little was known about hemophilia at this point in time, there was some literature, especially in German, describing the disorder.

Prince Leopold exhibited the typical symptoms of severe hemophilia. He bled profusely when he lost his milk teeth and he had joint symptoms similar to those of arthritis. In a letter to a relative when Leopold was

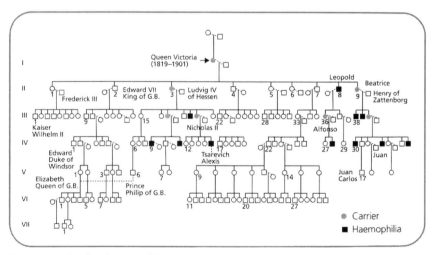

Figure 1.2 The family tree of Queen Victoria.

6 years old, his mother described for the first time that her son appeared to suffer from a serious illness for which there was no treatment and from which he probably would not recover. However, she gave no name to the illness. At the age of 8 years, he began to use knee pads as a result of repeated knee hemorrhages. Family photographs often portray him standing with one knee bent.

Leopold's upbringing came to be largely dominated by his bleeding disease, and his history shows many characteristics typical in families of hemophiliacs. Especially after the death of Prince Albert, Queen Victoria became extremely overprotective of her son, wanting him constantly by her side under the pretext that he required extra protection. This strongly inhibited Leopold's normal adolescent development and made him isolated and insecure. However, he finally succeeded in studying at Oxford where he achieved excellent study results. Leopold's history shows us that despite serious health problems and unsatisfactory relations with his mother who sought to control his life even as an adult, even exploiting his illness for this purpose, it is possible to succeed in acquiring an education and making a life of one's own! It pays to be stubborn! Similar patterns can be seen in the families of many hemophiliacs.

Leopold eventually married Helen Waldeck-Pymont, and they had two children, a daughter and a son who was born after the death of Leopold. The daughter had three children, a daughter, a son with hemophilia, and another son who died at the age of 6 months. Thus, the family exhibited the typical hereditary pattern in hemophilia—a daughter of a hemophilia patient who is a carrier of the disease but is not herself ill. She can, however, transmit the predisposition to a son, with a 50% risk that he will develop hemophilia.

The son, who did not have hemophilia, was made Duke of Sachsen-Coburg by Queen Victoria. Her husband Prince Albert was a descendent of this family. The young Charles-Edward, later Karl-Edvard, was then forced to move to Germany with his mother and sister in 1899. He married Victoria Adelheid in 1904 and had five children, the second of whom was Sibylla, born in 1908, who later married the Swedish Crown Prince Gustav Adolf, father of the present king, Carl Gustaf.

Another example of hemophilia, also described in the literature, is that of the last tsarevich Alexis of Russia who inherited his hemophilia from Princess Alexandra, daughter of Queen Victoria's daughter Alice who was married to a German archduke. The fate of the Russian tsar family is described in the book *Nicholas and Alexandra* by Robert Massie, published in 1967 [13]. The book is based on diaries and letters which the author found in various libraries and archives, mainly correspondence between tsar Nicholas and his wife Alexandra, but also correspondence between other relatives. The detailed descriptions of life at the imperial court, which include descriptions of tsarevich Alexis and his bleeding disorder, give a graphic and relevant picture of the situation in the life of a family where one of the members has hemophilia (Fig. 1.3).

Alexis was born in August 1904; when he was 6 weeks old, he developed an umbilical bleed for no known reason. It lasted for 3 days. This episode aroused his parents' suspicion that he possibly had hemophilia. His mother, Alexandra, had, herself, a brother with hemophilia who died at the age of 3 years. She also had several other relatives with

Figure 1.3 Tsarevich Alexis with his left foot on a stool because of joint changes in his hip and his knee, due to repeated joint bleedings, typical for hemophilia.

hemophilia—all related to Queen Victoria of England. When Alexis began to creep and stand up, he developed large swellings following minor blows, and the diagnosis of hemophilia was confirmed. In the book *Nicholas and Alexandra*, it is possible to follow the life of a child suffering from severe hemophilia. The mother's constant anxiety is compared to that of a soldier who, after long periods of stress, experiences total fatigue, resulting in complete emotional exhaustion. The soldier has to be removed from the front and ordered rest. In contrast to the soldier, there is no relief for the mother of a hemophilia boy and no rest from the ever-present battlefield, except for short periods when the child is asleep. The picture painted is a familiar one for the parents of most hemophilia patients, and the book is to be recommended for those interested in learning more about the disease.

It has even been suggested that "the nursery was the center of all Russia's troubles" and was the final cause of the Russian revolution and the fall of the Russian Empire. Like her grandmother, Queen Victoria of England, the Russian Empress refused to accept her fate and complained of the physicians' incompetence. She turned to religion for comfort, which brought her into contact with Rasputin, an itinerant monk from Siberia, who was believed to perform miracles. Rasputin came to exert great influence over the Empress, and the Russian government thus enabled him to precipitate the fall of the Romanov family and the Russian Empire.

1.2 THE TREATMENT OF HEMOPHILIA

During the 1800s and early 1900s there was no adequate treatment of bleeding in hemophilia. The disease was surrounded by mystery. Certain families were known to be prone to it, but possible heredity was shrouded in shadow. Examples of this are the sporadic cases showing up in the royal families of Central Europe during the 1800s. Thus, the Spanish royal family, as well as several archduke families in Germany, had members with hemophilia. They all showed an inheritance from Queen Victoria through her daughters [14]. Recently, it was discovered that the hemophilia trait of the European royal families was of type B, meaning that the patients were lacking factor IX (FIX) [15].

During the early years of the 1900s, blood and plasma transfusions began to be used. With the discovery of the blood groups by

K. Landsteiner in 1900, transfusions became safer. Not until in 1914 when sodium citrate, which prevents the coagulation of the blood, was introduced, it became possible to tap blood from a donor to a receiver through a vessel outside the body. Normally blood clots immediately when removed from the body, which prevents its use for transfusion into another human.

Blood and, in some cases, plasma (the blood liquid after the red blood cells have been removed) transfusions were introduced in Sweden by Dr. Erik Sköld who started the transfusion unit at St. Erik's Hospital in Stockholm in 1934. In his doctoral dissertation in 1944, he reported results from 15 hemophilia patients who had been given a total of 200 transfusions. In all cases, the coagulation time was notably shortened after transfusion. It was more remarkable that he reported the effect to remain for 20−30 days after treatment [16].

In the 1940s, two types of hemophilia were discovered, hemophilia A and hemophilia B. Both types have the same symptoms and course but are caused by the lack of two entirely different proteins, factor VIII (FVIII), in hemophilia A and FIX, in hemophilia B. The content of the two hemophilia factors in the blood is very low. It is therefore impossible to add enough to ensure full effect in cases of serious hemorrhages solely with blood or plasma. Treatment with transfusion does not have sufficient effect on surgical procedures. In an article from 1938, H.O. Mertz & L.T. Meits reported that 25% of hemophilia patients died in connection with surgery [17].

In this context an article published in 1909 by a Swedish doctor at the Malmö General Hospital, Otto Löfberg, is of interest. He described a case of a 26-year-old man requiring acute surgery. The patient had hemophilia but did not mention this before the operation. The next day the patient died of severe bleeding. It is, therefore, no wonder that surgery was advised against, unless absolutely necessary, for hemophilia patients [18].

In teeth extraction, which is the most common form of surgery in hemophilia patients, Erik Sköld used repeated plasma transfusions as well as a very meticulous technique to facilitate the coagulation of the blood locally. To relieve grave joint deformities, he successfully used gradual traction in a plaster cast under the cover of repeated transfusions. This type of treatment helped a few severely handicapped hemophilia patients to such a degree that they were able to abandon their wheelchairs and move around with the help of crutches or a stick [16].

Erik Sköld also compiled one of the first registers in the world of known hemophilia patients and their families This has been a great help in later work with hemophilia in Sweden. With the help of his register, Sköld was already able to describe the considerable variation in the symptoms and bleeding frequency in hemophilia patients. The cause of this variation is still unclear [16].

The major breakthrough in the treatment of hemophilia in Sweden came in the 1950s. Erik Jorpes was at that time head of the Centre for Medical Biochemistry at Karolinska Institutet in Stockholm. He came to Sweden as a refugee from Finland in 1919. In his department he gathered a group of young biochemists, mostly medical students. His interest in heparin inspired several of them to become interested in coagulation in general. Thus, in his department, fibrinogen, plasminogen, and prothrombin were purified. Several of his collaborators who worked on these proteins later became professors in various parts of the world, e.g., Per Wallén, Staffan Magnusson, Sven Gardell, Agnes Henschen, Birger, and Margareta Blombäck are among others.

Erik Jorpes himself worked on purifying heparin for clinical use. He was encouraged in this work by Clarence Crafoord, Professor of Surgery at Sabbatsberg Hospital in Stockholm, who needed heparin to treat thrombotic complications related to surgical procedures. He also needed heparin to enable him to perform open-heart surgery, which he had started to do in the 1940s.

Birger Blombäck joined the group in 1951 and was given the task of purifying fibrinogen. Pure fibrinogen was necessary for Erik Jorpes' development of a method to determine heparin activity in the blood. Together with his wife, Margareta, Birger Blombäck developed a purification method based on Cohn's plasma fractionation technique [19].

1.2.1 Fraction 1−0

The Blombäcks used Cohn's technique to separate blood plasma proteins into different fractions according to the qualities in the precipitation in various concentrations of alcohol. If Fraction 1 containing fibrinogen was extracted in the presence of glycine, most of the impurities were removed and a stable fibrinogen fraction was obtained. This was called Fraction 1−0 [20]. It was known that Cohn's fraction contained an "antihemophilia component."

Inga Marie Nilsson, who was a young medical doctor, at that time, at the Clinic of Internal Medicine at the Malmö General Hospital, headed by Professor Jan Waldenström, came to Erik Jorpes' laboratory as a visiting scientist in the middle of the 1950s to learn how to determine the heparin content in blood. She needed this for her dissertation work at the hospital in Malmö, part of Lund University, and collaboration with Birger and Margareta Blombäck was initiated. Inga Marie Nilsson treated hemophilia patients in Malmö and was thus familiar with laboratory methods for determining different coagulation factors. This collaboration led them to note that Fraction 1−0 contained a fairly high concentration of the hemophilia A factor. This discovery inspired Birger Blombäck and his colleagues to start producing Fraction 1−0 in the basement at Karolinska Institutet (Fig. 1.4). The work was associated with numerous problems that are needed to be solved before Fraction 1−0 could be administered to patients. In her personal notes, Margareta Blombäck declares that the work of producing Fraction 1−0 would have been impossible without Erik Jorpes' research resources and his contacts with Erik Sköld.

Figure 1.4 Margareta and Birger Blombäck in their laboratory at the Karolinska Institutet, Stockholm, Sweden, in the 1950s with one of the first bottle of Fraction 1−0.

1.2.2 The clinical use of Fraction 1−0

The first patient treated with Fraction 1−0 was a girl in Malmö with von Willebrand's disease. She was given Fraction 1−0 in connection with life-threatening menstrual bleeding, and it had an immediate effect. Fraction 1−0 was found to contain both FVIII, which is lacking in hemophilia A, and the von Willebrand factor. In close collaboration, Blombäcks' group in Stockholm and Inga Marie Nilsson in Malmö found that an "anti-bleeding factor" lacking in patients with von Willebrand's disease was present in factor 1−0 but not in pure FVIII or pure fibrinogen. This was named the von Willebrand factor and normalized the prolonged bleeding time in von Willebrand patients [21,22].

After the encouraging initial results with the administration of Fraction 1−0 to hemophilia patients, there followed a period, at the end of the 1950s and the beginning of the 1960s, with intensive treatment for hemophilia A patients in both Stockholm and Malmö. In Malmö, Inga Marie Nilsson and Åke Ahlberg, an orthopedic surgeon, started a project with the aim of correcting the pronounced joint defects found in most patients with severe hemophilia A at that time. For the first time, this advanced and complicated surgery could be performed in patients with severe hemophilia without the risk of hemorrhages. As part of this work and with the help of Erik Sköld's register, Åke Ahlberg examined almost all the hemophilia patients in Sweden and classified their joint damage [23].

Using Cohn's plasma-fractionation technique, a concentrate similar to the Swedish factor 1−0 was produced in England and in France. At the beginning of 1960s, the Blombäcks introduced the production of Fraction 1−0 in Australia, where it was used for a long time to come.

In 1965, Judith Pool in California published her observation that a large proportion of FVIII activity was recovered in the fibrinogen precipitate if frozen plasma was thawed slowly. This so-called cryoprecipitate could be centrifuged down and frozen for storage. This technique was frequently used and is still used for the treatment of hemophilia in many countries [24].

1.2.3 Prophylaxis in severe hemophilia

In his orthopedic review of hemophilia patients, Åke Ahlberg established that patients with moderate hemophilia and an FVIII concentration in the blood around 5% had considerably fewer joint damages than those with severe hemophilia (<1% FVIII in the blood). This study of 100 hemophilia patients showed that those with severe hemophilia had severe joint damage already at the age of 5–9 years, whereas those with moderate hemophilia seldom developed severe joint damage before the age of 20 years. This eventually led to the introduction of a regular administration of Fraction 1–0 to patients with severe hemophilia A. According to Inga Marie Nilsson, this should enable severe hemophilia A to be transformed into a moderate form. Sweden was the first country to introduce the regular use of a procoagulant concentrate without the presence of ongoing bleeding in the late 1950s. This is now a standard practice in the Western world [25].

At this time, the availability of FVIII concentrate was limited, and thus, only low concentrations of FVIII (10–20 IU/kg bodyweight) were administered once a month (1958–64). From 1964 to 1970, as the availability of Fraction 1–0 increased, the same dose was administered twice a month. With this dosage, most of the patients had an FVIII level in the plasma of 1% or lower after only a few days. Even with this relatively low dosage, which did not achieve the aim of transforming severe hemophiliacs into moderate ones, the number of hemorrhages decreased. Furthermore, those that did occur were most often milder than usual. In long-term follow-ups, it was seen that patients who had started prophylaxis before any marked joint changes had developed only had minor ones. At the same time, the number of days in hospital had decreased markedly [25].

During the following years, dosage was increased (25–40 IU FVIII/kg bodyweight, three times a week). With this dosage, the yearly consumption of FVIII reached 3500–5000 IU/kg. In patients who had started treatment at the age of 1–2 years, a bleeding frequency of 0.2 per year and largely undamaged joints were seen on X-rays. Surprisingly enough, there seems to be no clear correlation between dosage and achieved improvement of joint status. However, it was clear that the degree of damage was dependent on the number of joint hemorrhages the patients had had [26].

It would be interesting to see whether it would be possible to achieve a prophylactic effect regarding the development of joint damage even with FVIII/IX concentrations in the blood of 1% or less. It has long been known that the majority of coagulation factors are found not only in the blood but also in the tissues outside the vessels such as the lymph and in the synovial liquid [27,28]. Much points to a basal hemostasis taking place outside the vessels. The presence of activation peptides originating from activation of coagulating factors X, IX, prothrombin (F1 + 2), and fibrinogen in the blood of normal individuals supports the existence of such basal extravascular coagulation [28].

Thus, small amounts of several coagulation factors seem to be able to pass through an undamaged vessel wall and be at hand to immediately stop minor bleeding from small injuries in the tissues surrounding the joints. This normal basal hemostasis does not take place in hemophilia patients who lack the proteins FVIII/FIX, thus leading to the development of larger hemorrhages. Such micro-hemorrhages are most likely the cause of joint damage in hemophilia patients. It is possible that small amounts of FVIII/FIX, which can also pass through a nondamaged vessel wall, are sufficient to stop bleeding at an early stage and thus have a prophylactic effect on the development of joint hemorrhages in hemophilia patients [29].

REFERENCES

[1] Otto JC. An account of an hemorrhagic disposition existing in certain families. Med Repository 1803;6:1—4.
[2] Hay J. Account of a remarkable haemorrhagic disposition existing in many individuals of the same family. N Engl J Med Surg 1813;2:221—5.
[3] Rosner F. Hemophilia in the Talmud and Rabbinic writings. Ann Intern Med 1969;70:833—7.
[4] Maimonides M. Laws of circumcision, book of adoration (Sefer Ahavah). Code of Maimonides (Mishneh Torah). Jerusalem: Pardes Publishers; 1957, Chapter 1, paragraph 18.
[5] Hopff F. Über die Haemophilie oder die erbliche Anlage zu tödlichen Blutungen (Inaug.-Abhandlung). Würzburg: Carl Wilhelm Becker; 1828.
[6] W. Bulloch, P. Fildes, Treasury of human inheritance, Parts V and VI, Section XIVa, Haemophilia. Also published as Eugenics Laboratory Memoirs XII, Francis Galton Laboratory for National Eugenics, University of London (UCH). London: Dulau and Co; 1911.
[7] Manteuffel Z. Bemerkungen zur Blutstillung bei Hämophilie. Dtsch Med Wochenschr 1893;19:665—7.
[8] Bizzozero J. Bizzozero's new corpuscle. Lancet I 1882;446.
[9] Addis T. The pathogenesis of hereditary haemophilia. J Pathol Bacteriol 1911;15:427—53.
[10] Lenggenhagen K. Die Lösung des Hämophilen Blutungs und Gerinnungrätsels. Mitt Grenzgeb Med Chir 1936;44:425—39.
[11] Brinkhous KM. A study of the clotting defect in hemophilia: the delayed formation of thrombin. Am J Med Sci 1939;198:509—16.
[12] Zeepvat C. Prince Leopold: the untold story of Queen Victoria's youngest son. Guildford: Sutton Publishing Limited; 1998.
[13] Massie RK. Nicholas and Alexandra. New York: Dell Publishing Co; 1967.
[14] Ojeda-Thies C, Rodriguez-Merchan EC. Historical and political implications of haemophilia in the Spanish royal family. Haemophilia 2003;9:153—6.

[15] Lannoy N, Hermans C. The "royal disease"—haemophilia A or B? A hematological mystery is finally solved. Haemophilia 2010;16:843—7.
[16] Sköld E. On haemophilia in Sweden and its treatment by blood transfusion. Stockholm, PA: Norstedt & Söner; 1944.
[17] Mertz HO, Meiks LT. The haemophilia as a surgical risk: report of a case of nephrectomy with death. Urol Cutan Rev 1938;42:557—62.
[18] Löfberg O. Kirurgiskt ingrepp på hemofil individ med dödlig utgång. Hygiea 1909;71:380—5.
[19] Blombäck B, Blombäck M. Purification of human and bovine fibrinogen. Arkiv Kemi 1956;10:415—53.
[20] Blombäck M. Purification of antihemophilic globulin. I. Some studies on the stability of the antihemophilic globulin activity in fraction I-0 and a method for its partial separation from fibrinogen. Arkiv Kemi 1958;12:387—96.
[21] Nilsson IM, Blombäck M, von Francken I. On an inherited autosomal hemorrhagic diathesis with antihemophilic globulin (AHG). Deficiency and prolonged bleeding time. Acta Med Scand 1957;159:35—57.
[22] Nilsson IM, Blombäck M, Jorpes E, Blombäck B, Johansson S-AV. Willebrand's disease and its correction with human plasma fraction I-0. Acta Med Scand 1957;159:179—88.
[23] Ahlberg Å. Haemophilia in Sweden VII. Incidence, treatment and prophylaxis of arthropathy and other musculo-skeletal manifestations of haemophilia A and B. Copenhagen: Munksgaard; 1965.
[24] Pool JD, Hershgold EJ, Pappenhagen AR. High-potency antihaemophilic factor concentrate prepared from cryoglobulin precipitate. Nature 1964;203:312—13.
[25] Nilsson IM, Hedner U, Ahlberg Å. Haemophilia prophylaxis in Sweden. Acta Paediatr Scand 1976;65:129—35.
[26] Nilsson IM, Berntorp E, Löfqvist T, Pettersson H. Twenty-five years' experience of prophylactic treatment in severe haemophilia A and B. J Intern Med 1992;232:25—32.
[27] Le DT, Borgs P, Toneff TW, Witte MH, Rapaport SI. Haemostatic factors in rabbit limb lymph: relationship to mechanisms regulating extravascular coagulation. Am J Physiol 1998;274:H769—76.
[28] Miller GJ, Howarth DJ, Attfield JC, Cooke CJ, Nanjee MN, Olszewski WL, et al. Haemostatic factors in human peripheral afferent lymph. Thromb Haemost 2000;83:427—32.
[29] Hedner U. Potential role of recombinant factor FVIIa in prophylaxis in severe haemophilia with inhibitors. J Thromb Haemost 2006;4:2498—500.

CHAPTER TWO

My Encounter With Hemophilia (1959−82)

Contents

2.1 THE COAGULATION LABORATORY IN MALMÖ

In the summer of 1959, the first summer of my medical studies, I worked as an unpaid laboratory technician in the coagulation laboratory at Malmö General Hospital (MAS). I wanted somewhere to work in the summer vacation, and my father, who was the hospital pharmacist, suggested that I ask Inga Marie Nilsson if she had a place for me. He was fascinated by her pioneering work with hemophilia patients. During that summer, I learned to take blood samples and perform various coagulation analyzes. Inga Marie Nilsson urged me to copy down all method descriptions by hand so that I would learn the principle of coagulation analysis (Fig. 2.1).

The hemophiliacs, who were inpatients at the medical clinic at MAS due to hemorrhages, spent a lot of time in the basement where the laboratory was situated if they were well enough to move around. Like the rest of the staff, I got to know these hemophilia boys extremely well. Their mothers also became good friends with everyone. A strong family bond and family feeling developed between patients and staff. I, myself, became good friends with the laboratory staff, a friendship that has lasted over the years.

Continued contact and vacation work, always unpaid, during the following years, enabled me to maintain contact with the coagulation unit in Malmö during the remaining years of my studies. Through Inga

Figure 2.1 My father, Eric Flodmark, to the left, my supervisor, Inga Marie Nilsson in the middle, and myself to the right at my doctor promotion in May 1974.

Marie Nilsson, it was made possible for me to work with John Sjöquist at the medical—chemical institution in Lund parallel to my studies. There I learned the basic techniques within protein chemistry. This also eventually led to several years' work as an assistant at the same institution.

During my years at the coagulation laboratory in Malmö, I was, among other things, given the task of developing new analysis techniques. This was naturally an efficient way for Inga Marie Nilsson to involve me in her upcoming research work. It also gave me a solid knowledge of analytic methods in the area of coagulation, something that has been of invaluable help to me during later years, not least in my work of developing FVII.

In 1972, I applied for, and was given, a newly created residency at the coagulation laboratory in Malmö. The new position meant that coagulation activity became an acknowledged specialty and the clinical work with hemophilia patients could be extended. Moreover, investigatory work, aimed at the diagnosis of hemorrhagic and thrombotic diseases, developed and eventually the name of "The clinic for hemostatic disorders" was introduced.

At this time, Fraction 1—0 treatment was well established for hemophilia A patients in Sweden (see Chapter 1: Classical Bleeding Disease (Hemophilia)). In the 1970s "prothrombin complex concentrate" (PCC) had also been introduced for hemophilia B patients. These concentrates were produced from plasma after the FVIII fractions had been removed, and they contained, as well as FIX, many other blood proteins. They fulfilled the requirements for the treatment of acute hemorrhages. Even major orthopedic surgery could be performed in hemophilia B patients with the protection of these concentrates [1].

However, the greatest problem at this time was the treatment of patients who had developed antibodies to the factor they were lacking. There were six such patients in Sweden. It is not an impressive number, but these patients kept us constantly occupied. They could not be treated with concentrates containing hemophilia A or hemophilia B factors as their antibodies immediately neutralized the effect of these factors. Several complicated methods had been developed to help patients with inhibitors. The great problem was that in most cases, the treatment had to start by lowering the concentrations of antibodies in the blood. To do this, the blood had to be taken out of the body and led through a cylinder filled with a sterile material on which the antibodies were adsorbed. In this way the amount of antibodies in the patient's blood was decreased, and the blood could then be led back to the patient's blood vessels (the so-called extracorporal circulation that is also used in heart surgery and the treatment of renal failure by hemodialysis) [2,3].

This is a complicated procedure and had to be followed by treatment with large amounts of FVIII or FIX preparations. Besides this, chemotherapy is required to reduce antibody-forming stimulation by the added concentrates. Such treatment could not be given in cases of the most common hemorrhages, that is to say, joint hemorrhages. It was, however, used in cases of vital surgery or life-threatening hemorrhages. Malmö had positive experience of this complicated technique, partly due to Inga Marie Nilsson's work. She developed a special method by which the content of antibodies in a patient's blood and the ability of the used concentrate of FVIII/FIX to neutralize the patient-specific antibody could be determined. This method gave us the possibility to determine the exact amount of factor concentrate we needed to add to achieve hemostasis in the patient in question [2,3].

The treatment of mild-to-moderate bleeding in hemophilia patients with antibodies against the factors they were lacking was, however, suboptimal. These patients often spent much time in hospital in conjunction with joint bleeding.

2.2 HEMOPHILIA PATIENTS WITH INHIBITORS, AND A METHOD TO STOP BLEEDING IN THESE PATIENTS

During the 1970s great progress in purifying and structurally determining the coagulation proteins was made in the area of biochemistry.

The subject of coagulation began to be respected even by doctors and scientists in the laboratory disciplines. However, this entailed certain risks. In their enthusiasm to introduce "real" methods that determined the actual protein, they tended to forget the function in the coagulation process. Consequently, this could have been catastrophic in the adjustment and follow-up of prophylactic thrombosis treatment with the antithrombin (AP) Waran. Patients treated with Waran have, namely, a normal content of the protein prothrombin, although in a changed form which is not active, and therefore the protein level in the plasma does not reflect the content of the changed inactive protein, i.e., the Waran effect [4].

New protein purification techniques (ion exchange chromatography and Sephadex fractioning, the latter developed by Pharmacia, Uppsala), had been introduced, which facilitated the work of producing pure proteins which could be structurally determined. Much of the characterization of coagulation proteins was done in Earl Davie's laboratory, University of Washington, Seattle, USA. This increased knowledge helped the area of coagulation to develop from a more or less suspicious discipline of "activities" to "genuine" protein chemistry. Against this background, the suboptimal treatment of inhibitor patients prevailing in the 1970s seemed unacceptable to me.

At this time, great effort was made to find an FVIII-bypassing therapy that was independent of the presence of FVIII and thus also independent of the presence of antibodies to FVIII. In the early 1970s, it had been observed that certain so-called FIX concentrates clotted in the test tube. They were called "auto-FIX" and seemed to prevent bleeding in hemophilia A patients with antibodies. The FIX concentrates contained a mixture of all vitamin K-dependent coagulation factors, in both inactive and active forms. This led to the conclusion that if a hemophilia A patient was given already activated vitamin K-dependent coagulation factors, these might possibly activate coagulation without the assistance of FVIII/FIX. Especially FIXa and FXa were discussed as the hemostatic factors in this context. These concentrates were later called "activated prothrombin complex concentrates," abbreviated aPCC, and they are still in common use in the treatment for hemophilia A patients with FVIII inhibitors. As they were initially intended for the treatment of hemophilia B patients, the preparation was primarily directed at obtaining a high content of FIX [5].

However, the degree of purification was generally low with a large content of plasma proteins other than the vitamin K-dependent coagulation factors. This type of concentrate (aPCC, FEIBA) has shown an efficacy of around 50%–70% in joint and muscle bleedings [6–8]. There was no consensus as to their efficacy in severe hemorrhages or in surgery. Also, there were reports of side effects in the form of blood clots in both arteries and veins (thromboembolic side effects). These complications include the development of clots in the small arteries of the leg, acute heart failure, and the so-called disseminated intravascular coagulation (the formation of blood clots in the small arteries of several of the body organs) [9–13]. The combination of an efficacy not exceeding 65% in controlled studies and reports of thromboembolic side effects made the use of these concentrates less attractive.

The use of aPCC, like other plasma-derived preparations, also had the potential risk of transferring blood-carried pathogenic elements, e.g., virus (HIV, West Nile Virus, hepatitis, etc.), bacteria, and prions (the cause of so-called Mad Cow Disease). Therefore, this did not seem to be a good solution for the patients. The need for an improved treatment for hemophilia patients with inhibitors was obvious.

Based on the discussions on the risk and benefit of the aPCCs and the lack of any solid data regarding factors that might be responsible for any side effects and/or benefits, I thought it would be relevant to study different aPCCs and their ability to induce a systemic activation of the coagulation process. The first goal in my work on improving the treatment options for hemophilia patients with inhibitors was to find out how to get rid of the factors in the aPCC preparation, which resulted in a general activation of the coagulation system. I used an experimental model established at Malmö General Hospital. In this model, dogs were injected with an aPCC preparation, which caused changes pointing to a general activation of the coagulation system. [14].

In a further experiment I was able to demonstrate that the addition of antithrombin and heparin minimized these laboratory changes [15]. The results were first made public in 1977, at a meeting of the "International Committee on Thrombosis and Haemostasis, held before the "Fifth International Congress on Thrombosis and Haemostasis" in Philadelphia on June 26–July 2, 1977. At this meeting, a working group with focus on "Factor IX complex and Factor IXa" had heated discussions as to which factor had possible hemostatic effects and which initiated the general activation of the coagulation system and thus causing thromboembolic side effects [16].

Next step in my search for a factor with a hemostatic effect but not inhibited by heparin and antithrombin ended up with FVIIa. Activated factor VII (FVIIa) had been demonstrated to lack enzymatic activity unless it had formed a complex with the so-called "tissue factor" (TF). As early as 1972, B. Østerrud and coworkers had made it known that the presence of TF highly enhanced the enzymatic activity of FVII/FVIIa [17]. Thus, it could be hypothesized that injected factor VIIa, which was not in itself active, might find its way to the TF exposed at the site of vascular injury. There it might form a complex with TF and initiate local hemostasis [18].

At this time, results were published from the first patients who had received the so-called "Auto-IX-concentrate." From the tables and figures in the article [19], I could see that the concentrate used was especially rich in FVII. Of all the factors tested, the plasma levels of FVII after injection of that special batch of "Auto-IX-concentrate" showed the most marked increase that suggested that FVIIa might be an attractive candidate for further study and a potential treatment possibility.

Although the importance of FVII in the initial phase of hemostasis had been pointed out earlier, the injection of extra factor VII had only been considered in connection with liver diseases. The hemostatic effect of aPCC was often attributed to the presence of FIXa and FXa. It was therefore important for me in the 1970s to find out the extent to which FVIIa, injected as a pure product without the addition of other activated coagulation proteins, would be able to stop bleeding in patients. At this point, in the middle of the 1970s, I had many discussions with Harold Roberts, the cochairman of the working group for the clinical use of FIX concentrate within the "International Society of Thrombosis and Haemostasis" (ISTH) and also with Earl Davie whom I met at the Lindeström/Lang conference in 1975.

My idea of using FVIIa in the treatment for hemophilia patients with inhibitors was met with skepticism. It was pointed out that hemophilia patients have normal concentrations of FVII, so why should they be helped by additional FVII?

2.3 PLASMA FVIIa FOR HEMOPHILIA PATIENTS

In August 1978, I came to Earl Davie's laboratory in Seattle, USA, as a visiting scientist where I came to share an office with Walter Kisiel.

He was, at that time, working on the purification of FVII from human plasma. I started to discuss with him the possibility of purifying FVII to test in animals and later in humans. He stressed that it was difficult to purify FVII from human plasma. In the meanwhile, I worked on an entirely different project in Earl Davie's laboratory and followed the ongoing discussions about the active components in "auto-IX concentrate." To my delight, in an article by U. Seligsohn et al. from Sam Rapaport's research team, I found results that I thought supported my idea of using FVII in the treatment of hemophilia [20].

In our shared office in Earl Davie's laboratory, we gradually began to discuss the treatment of hemophilia and Walter Kisiel became more and more interested in the disease. He finally came up with the idea of spending some time at the hemophilia clinic in Malmö. Together, we planned a research project about FIX and variants of the FIX molecule. We had worked with a similar project earlier in Malmö. In this project, FIX variants with different adsorption capacities to specific FIX antibodies had been analyzed. I sent our application to the Swedish Medical Research Council and, at the same time, applied for a fellowship for Walter Kisiel, so that he could spend a year at our clinic. The application was approved, and in July 1980, Walter Kisiel and his family came to Malmö.

An interesting time followed, with Walter Kisiel working on the purification of FVII and FIX from human plasma. At the same time, he learned more about hemophilia and the daily problems these patients faced, especially those with inhibitors. This helped him to finally understand why I was so obsessed with the idea of finding a treatment for these patients as effective as that given to patients without inhibitors. My vision in the late 1970s was to find a treatment which could be used at home and which was also effective in major surgery. It had previously been almost impossible to perform these operations on inhibitor patients.

After contact with the Swedish Health Authorities, I started to work with a formula of pure activated FVII, not previously produced and prepared from human plasma by Walter Kisiel in our laboratory at the University Hospital in Malmö. I followed the guidelines given to me by a personal contact with the Health Authorities. We also applied for and were granted permission by the Ethical Committee at Lund University to use our preparation in the treatment for patients.

2.4 TREATMENT FOR THE FIRST PATIENTS WITH FVIIa

In March 1981, we had tested our purified FVIIa in the same dog model that I used previously to test the aPCC and found no signs of a systemic activation of the coagulation system. Pure FVIIa, therefore, should not give rise to the side effects that were seen with the use of aPCC. During the discussion at a congress arranged by Immuno AG in Rome on March 31, 1981, I presented our results mentioning that we intended to treat a hemophilia patient with inhibitors as soon as anyone presented with an acute bleeding in our clinic in Malmö. Accordingly, the first patient was treated with plasma-derived FVIIa (pd-FVIIa) on April 24, 1981.

The patient had a muscle hemorrhage and despite such bleedings being difficult to assess, it was clear that the patient recovered much quicker than during similar earlier bleeding episodes. He was able to leave the hospital the day after treatment. Unfortunately, the next day, he happily went sailing with his father when a new hemorrhage occurred. He was then given further treatment with pd-FVIIa.

The tension was high when this first dose of pure plasma-prepared FVIIa was injected into a hemophilia patient with an FVIII inhibitor. The whole idea of testing the effect of factor VIIa in this situation was met with extreme suspicion in my immediate surroundings and, in retrospect, I hardly understand how I had the courage to carry out the treatment. I realized that if anything went wrong, I would be met with a massive reaction of the type "I told you so" accompanied by a large measure of malicious glee.

Another patient was treated a year later when he lost a primary molar (Fig. 2.2). In conjunction with similar episodes of tooth loss, this patient had previously been treated at the hospital in Malmö with exchange transfusion and large doses of FVIII involving long hospital stays. The treatment with FVIIa was the result of an agreement between the boy's mother and myself that we should try FVIIa when he next loses a tooth. We agreed that his mother would contact me when the next tooth was loosening and I would take FVIIa with me to the hospital in his hometown. One day she called and I took the night train that arrived to his hometown at 5 o'clock in the morning.

After breakfast with one of my colleagues, head of the hospital's transfusion center, I went to the ward in the children's clinic where my patient

Figure 2.2 The second patient treated with plasma-derived FVIIa (pd-FVIIa). He has hemophilia with typical joint defects in both his knees caused by repeated joint bleedings. Due to contractions in both his knee joints, he cannot stretch his legs and thus has difficulties in walking.

was admitted. In the vestibule on my way there, I met the boy's mother who was extremely agitated. The dentist had just been in to see the patient and extracted the loose tooth that had still been attached by one root. Both the mother and I knew that this was never done with a hemophilia patient. Instead the tooth should be gently coaxed out while using a hemostatic medicine. This is an excellent example of the need to centralize hemophilia care. The local dentist was not aware of the risks, having no experience of a hemophilia patient. It was obvious that the boy's mother was upset.

I was, of course, also worried as I knew that if the pd–FVIIa preparation did not work satisfactorily, I would need to take him to Malmö by helicopter and treat him with exchange plasma transfusion and large doses of FVIII as in previous similar situations. It was not an attractive prospect and obviously I felt nervous. We went in to the boy and I started to prepare the ampoules with FVIIa, which I had brought in a freezer bag, with dry ice, and injected the first dose. The boy's mother sat on the other side of the bed and we watched intensively the swab that had been placed on his gum where the tooth had been. The mother had already changed the swabs several times as they quickly became red due to bleeding. Soon after the first injection of FVIIa, she cried out "It's working! The swab isn't getting red so quickly." After carefully removing the swab, we saw that a dense, fine clot had been formed where the tooth had been and that there was no ongoing bleeding. We both experienced a great sense of relief. For me it was a clear "proof of principle" that extra injected purified FVIIa worked in cases of severe hemophilia with inhibitors.

The same patient was given the same treatment when he lost another tooth, and Walter Kisiel and I started to write an article that we sent to the journal *Blood*, where it was rejected. However, it was later accepted by the *Journal of Clinical Investigation* and published in June 1983 [18].

2.5 POTENTIAL DEVELOPMENT OF FVIIa TO BE USED IN HEMOPHILIA

To follow up the potential development of FVIIa to be used in the treatment of hemophilia, discussions were held between AB Kabi Vitrum in Stockholm, Walter Kisiel, and myself during the latter part of 1982. The discussions did not lead anywhere. The research directors thought that the whole thing was too risky. They were obviously advised by other hemophilia experts in the area not to cooperate with us on the project which they considered extremely hazardous.

In October 1981 I was invited to a meeting at Armour Pharmaceutical Company, Revlon Health Care Group, Tuckahoe, New York, USA, to discuss about the treatment for hemophilia patients with inhibitors. The meeting was arranged by a person from the Revlon Health Care Group, Europe, and I was asked to present my idea of using FVIIa for the treatment of bleeding in inhibitor patients. The conclusion of the discussion was that the idea seemed to be a "sound and exciting approach. It should be seriously discussed as a new product development" [21]. However, it was concluded to follow up on future data and progress. At this time only the possibility of producing pd-FVIIa was discussed, and nothing came out of the meeting.

REFERENCES

[1] Nilsson IM, Ahlberg Å, Björlin G. Clinical experience with a Swedish factor IX concentrate. Acta Med Scand 1971;190:257—66.
[2] Nilsson IM, Hedner U. Immunosuppressive treatment in haemophiliacs with inhibitors to Factor VIII and Factor IX. Scand J Haematol 1976;16:369—82.
[3] Nilsson IM, Jonsson S, Sundqvist SB, Ahlberg Å, Bergentz SE. A procedure for removing high titer antibodies by extracorporeal protein A-Sepharose adsorption in hemophilia: substitution therapy and surgery in a patient with hemophilia B and antibodies. Blood 1981;58:38—44.

[4] Niléhn JE, Ganrot PO. Plasma prothrombin during treatment with dicumarol I. Immunochemical determination of its concentration in plasma. Scand J Clin Lab Investig 1968;22:17—22.

[5] Fekete LF, Holst SL, Peetom F, and deVeber LL. "Auto" factor IX concentrate: a new therapeutic approach to treatment of haemophilia A patients with inhibitors. XIV Congresso Internacional de Hamatologia, San Paolo, Brasil, 26—21 July 1972, Abstract 295.

[6] Sjamsoedin LJM, Heijnen L, Mauser-Bunschoten EP, van Geijlswijk JL, van Houwelingen H, van Asten P, et al. The effect of activated prothrombin-complex concentrate (FEIBA) on joint and muscle bleeding in patients with hemophilia A and antibodies to factor VIII. N Engl J Med 1981;305:717—21.

[7] Lusher JM, Blatt PM, Penner JA, Aledort LM, Levine PH, White GC, et al. Autoplex versus Proplex: a controlled, double-blind study of effectiveness in acute hemarthroses in haemophiliacs with inhibitors to factor VIII. Blood 1983;62:1135—8.

[8] Young G, Shafer FE, Rojas P, Seremetis S. Single 270 µg/kg-dose rFVIIa vs. standard 90 µg/kg-dose rFVIIa and APCC for home treatment of joint bleeds in hemophilia patients with inhibitors: a randomized comparison. Haemophilia 2008;14:287—94.

[9] Cederbaum AI, Blatt PM, Roberts HR. Intravascular coagulation with use of human prothrombin complex concentrates. Ann Intern Med 1976;84:683—7.

[10] Fuerth JH, Mahrer P. Myocardial infarction after factor IX therapy. J Am Med Assoc 1981;245:1455—6.

[11] White GC, Roberts HR, Kingdon HS, Lundblad RL. Prothrombin complex concentrates: potentially thrombogenic materials and clues to the mechanism of thrombosis in vivo. Blood 1975;49:159—70.

[12] Lusher JM. Thrombogenicity associated with factor IX complex concentrates. Semin Hematol 1991;28:3—5.

[13] Blatt PM, Lundblad RL, Kingdon HS, McLean G, Roberts HR. Thrombogenic materials in prothrombin complex concentrates. Ann Intern Med 1974;81:766—70.

[14] Hedner U, Nilsson IM, Bergentz SE. Various prothrombin complex concentrates and their effect on coagulation and fibrinolysis in vivo. Thromb Haemostas 1976;35:386—95.

[15] Hedner U, Nilsson IM, Bergentz SE. Studies on thrombogenic activities in two prothrombin complex concentrates. Thromb Haemostas 1979;42:1022—32.

[16] Hultin M. Activated clotting factors in factor IX concentrates. Blood 1970;54:1028—38.

[17] Østerud B, Berre Å, Otnaess, Bjørklid E, Prydz H. Activation of the coagulation factor VII by tissue thromboplastin and calcium. Biochemistry 1972;11:2853—7.

[18] Hedner U, Kisiel W. Use of human factor VIIa in the treatment of two haemophilia A patients with high-titer inhibitors. J Clin Investig 1983;71:1836—41.

[19] Kurczynski EM, Penner JA. Activated prothrombin concentrate for patients with factor VIII inhibitors. N Engl J Med 1974;291:164—7.

[20] Seligsohn U, Kasper CK, Østerud B, Rapaport SI. Activated factor VII: presence in factor IX concentrates and persistence in the circulation after infusion. Blood 1979;53:828—37.

[21] Protocol from the meeting at Amour Pharmaceutical Company, Revlon Health Care Group, Tuckahoe, New York, USA, October 26, 1981.

CHAPTER THREE

The First Years at Novo Nordisk

Contents

3.1 RECRUITMENT BY NOVO NORDISK, 1983

At the beginning of the 1980s I had some contact with Novo Nordisk, then Novo Industri A/S, who, apart from being one of the leading producers of insulin, had manufactured heparin since the 1920s. During the 1970s the treatment of thrombotic diseases with heparin had changed to a certain extent. The so-called "low dose heparin treatment" was introduced to prevent the development of thrombosis [1]. Besides this, new discoveries about the chemical properties of heparin had led to the idea that it might be beneficial to split the heparin molecule into smaller parts and use those with lower molecular weight in the treatment of clotting diseases, both to prevent thrombosis and to treat clots which had already been formed [2,3].

Being aware of this, Novo had started a project with the aim of developing a "low molecular weight heparin" (LMWH). The first studies of LMWH were carried out on normal voluntary subjects at the University Hospital in Malmö. I was involved in this project together with David Bergqvist, who was an Associate Professor of Surgery at Malmö at that time. The continued development of Novo's LMWH, Tinzaparin later became my first project at Novo [4,5].

At this time, Novo had decided to expand their involvement in hemostasis, primarily in the field of thrombosis treatment. At the beginning of the 1980s, work to produce insulin with the help of gene therapy was going on. For this purpose, yeast cells were used, a technique that was well developed at the University of Washington in Seattle, USA. Novo had already established cooperation with yeast scientists at this university, largely because Niels Fiil, Head of Molecular Biology at Novo had spent some time at the University of Washington, working with the well-known yeast specialist Ben Hall. He also knew about Earl Davie's group at the Department of Biochemistry at the same university, which had considerable expertise in the area of hemostasis.

As Novo's research management knew that I had spent more than a year in Seattle with this specialized hemostasis group, I was asked if I would accompany a delegation from Novo and introduce them to Professor Earl Davie. I had naturally nothing against this as I still collaborated with Earl Davie's group and often had reason to go there.

The journey to Seattle took place in autumn 1982. The possibility was discussed of manufacturing a product to dissolve clots formed, in the coronary arteries and causing an acute myocardial infarction. However, to develop a product for use in this area, a research group specialized in thrombotic disease at Novo was required. In January 1983, Novo contacted me to ask if I would be interested in establishing such a research group. During a lunch in Malmö with Novo's Head of Research, Professor Ulrik Lassen, and Peter Tang, who was the Director of Research within the area of products and development and who became my future chief, I was given a detailed description of the new laboratory building in Bagsvaerd, outside Copenhagen. I was hesitant at first, for although the offer was tempting, I had no experience of the pharmaceutical industry. However, after taking the matter into consideration for 6 months, I accepted and started at Novo in September 1983.

I applied for leave of absence from my consultancy in Malmö and enthusiastically embarked on my new undertaking in Copenhagen. I was appointed as "senior chemist, responsible for clinical research, thrombosis" from September 15, 1983, onward. An additional clause in the contract of employment read, "if the company makes an economic gain from the employee's contribution or a contribution which in some other way is advantageous for the company, the employee will be awarded financial recompense in proportion to the importance and circumstances under which it is made." At the time I did not pay any attention to this nor have I so far been able to make use of it.

My first task at Novo was to set up a laboratory specialized in analyzing coagulation factors. I did this together with a skilled laboratory technician, Dorthe Winter, who I immediately was allowed to recruit. She was still, in 2014, working at Novo Nordisk. A laboratory with standardized methods was a necessity for pursuing research in this area. Dorthe Winter and I used the basic methods of laboratory analyses that I had learned during my time as a student working as a technician at the coagulation laboratory in Malmö.

After a short time I was given responsibility for the section at Novo working on producing plasmin (the most effective enzyme for the breakdown of fibrin and fibrin thrombi) from pig plasma. This plasmin, prepared from pig blood (Lysofibrin), had been used to dissolve thrombi in a few patients but was never developed into a registered drug. As a consultant to Novo, during my time in Malmö, I had treated at least one patient with Lysofibrin. The same section had the responsibility for the production of heparin, which Novo Industry had marketed since 1952.

3.2 HEMATOLOGY AT NOVO INDUSTRI DURING THE 1980s

3.2.1 The history of heparin and Novo Industri

Heparin and its anticoagulating effect were already known in the early 1920s. The discovery was made almost by chance and was the result of laboratory work done by a medical student, Jay McLean, in William H. Howell's laboratory at Johns Hopkins University in the United States. During his work to produce coagulation-promoting substances from various human organs, he found those with the opposite effect in the liver and heart. These were the first anticoagulant substances found in human tissue, and the discovery was presented in 1916 [6,7]. Howell's group then demonstrated anticoagulant substances, above all in the liver, which is why it was given the name "heparin" [8]. Howell presented a report on the chemistry and physiology of heparin in 1928 [9]. This heparin, however, could only be used for research on animals, as it had serious side effects such as fever and shock condition.

Continued work on heparin was conducted in the Connaught Laboratories at the University of Toronto, Canada, under the auspices of physiologist Charles Herman Best (1899−1978). Best's aim was to find a safe source of heparin and develop a purification procedure that would enable heparin to be given both to humans and to animals without serious side effects. The latter aim was reached in 1937 with the production of a purified form of heparin that became commercially accessible. The chemical structure of heparin was identified at the Connaught Laboratories in Toronto and by Erik Jorpes at Karolinska Institutet in Stockholm during the 1930s [10]. The first systematic studies of heparin as prophylaxis in connection with surgical procedures were carried out during 1935−36 by Clarence Crafoord, Professor of Surgery in Stockholm, and the results were published in 1937 and 1941 [11,12].

The history of heparin production was started in the Connaught Laboratories in Toronto and was strongly linked to insulin, which was produced in the same laboratories and partly by the same scientists. One of these was a Danish scientist, Albert Fisher, associated with the Carlsberg Foundation in Copenhagen. His main interest was the purification of insulin, but he also worked with the heparin group. Among other things, it was through him that a close collaboration was developed not only with Stockholm (through E. Jorpes and the Vitrum AB company) but also with Copenhagen. During the latter part of the 1940s, an improved technique for the purification of insulin was developed in Copenhagen, and Novo Industri A/S was offered the opportunity of taking over the production of heparin. From 1952 onward, heparin was marketed by Novo Industri A/S in Denmark.

Thanks to heparin, cardiac surgery, dialysis in cases of renal failure, and many other procedures were made possible. Access to heparin enabled Clarence Crafoord to perform the first heart operations using extracorporal circulation leading the patient's blood, after an injection of heparin, from the body to a heart−lung machine that temporarily replaced the patient's heart and lungs (and provided the blood with oxygen). The world's second open-heart surgery was carried out in 1954 with the cooperation of thorax surgeon Per Olsson, assisted by Margareta and Birger Blombäck who carried out the heparin analyses during the surgery [13]. Further development of cardiac surgery also became possible thanks to the access to heparin [14].

In cases of renal failure, it is necessary to lead the patient's blood out of the body to a machine that clears the blood from slag products. It is said that Dr. Willem Johan Kolff from Holland succeeded in traveling to Stockholm in the middle of World War II to obtain heparin from Erik Jorpes. He had developed a filtration method with the aid of a washing machine but needed heparin to lead the patient's blood out of the body. Dr. Kolff's work was the starting point for the development of modern hemodialysis for treating patients with chronic renal failure. This treatment is still dependent on access to heparin.

3.2.2 The development of modified heparins

During the 1970s, new methods were developed for the purification of complex carbohydrates. This made it possible to study the structure of heparin more closely. The new methods developed during the 1970s made it possible to separate fractions of heparin that reacted with different coagulation proteins [15]. As early as 1939 it had been seen that heparin was dependent on the presence of a plasma protein to function as an anti-coagulant [16]. This protein proved to be identical with the clotting inhibitor antithrombin [17].

The combination of heparin and antithrombin was later found to noticeably accelerate all the inhibitory activities of antithrombin targeted on several factors promoting clotting. Detailed studies of the interaction between heparin and antithrombin in the latter part of the 1970s led to the development of the so-called LMWHs.

3.2.3 The development of LMWH at Novo Industri

Novo had long-standing experience with the production of industrial enzymes and had developed heparinase, an enzyme that breaks down heparin into fragments with lower molecular weight than the standard heparin. It was, therefore, a natural step for Novo to use this enzyme for the production of an LMWH. This development started during 1978–82.

The further development of LMWH at Novo became my first hemostasis project when I went to work for Novo in the autumn of 1983. The relevant clinical advantages of the heparinase-produced LMWH, which was called Tinzaparin, and later Logiparin were stressed. At this stage, our group had access to two chemists for the first 2 years and four or five during the third year. We were faced with the task of starting a clinical investigation that is the most costly phase in all pharmaceutical production. Our first clinical study of Tinzaparin was carried out in patients who were to undergo some form of surgical procedure associated with an increased risk of thrombus formation. The study was to include 1290 patients that is not an impressive number compared with the number normally required to show a significant effect in similar patient groups. The number of patients was a shock to Novo, who had never before carried out such a large clinical study. It met with a marked resistance on account of the large costs involved and the low expectations in general on LMWHs at Novo. Expectations in other parts of the clinical world were, on the contrary, high. It turned out to be difficult or rather impossible to convince the management of Novo about the need for an improved heparin in the clinic. I started to learn that it was not easy to introduce a new area of therapy in a company that was, and still is, an insulin company.

We succeeded in carrying out two studies on Tinzaparin as prophylaxis against thrombosis, the second one including patients undergoing orthopedic surgery with a knee implant, a procedure with a high risk of thrombosis. These studies showed, for the first time, that an LMWH, i.e., Tinzaparin, only needed to be administered to the patient once a day to achieve a significant prophylactic effect [18]. The use of LMWHs was not accepted at this time as a treatment for existing blood clots. In these patients standard heparin in continuous intravenous infusion was administered. Tinzaparin proved to be the first LMWH shown to be effective even in the treatment of established thrombosis [19].This LMWH was characterized by an extremely small structural change compared to the original heparin molecule, which we considered to be an advantage.

According to a summary of our Tinzaparin project in 1991, it had taken 9 years to have it registered on the market (launched in Denmark in February 1991). About 15% of the money made by the sale of standard heparin had been invested in the LMWH project. In the summary we could establish that LMWH had taken over the market in several countries. Novo Nordisk also was the first producer of LMWH (Tinzaparin) to demonstrate that the use of LMWH meant a simpler and more efficient treatment of established thrombosis. In addition, our treatment study showed a lower cancer mortality among the patients who had been treated with Tinzaparin as compared with those who had been treated with standard heparin. As a result of these findings, FDA (Food and Drug Administration, responsible for approval of new drugs in the United States) recommended us to ask for approval of the indication for use of Tinzaparin in cancer patients. Despite this and the constantly increasing marked of LMWHs in several countries, Novo Nordisk later decided to sell off Tinzaparin to Leo Pharmaceuticals, who is currently successfully selling Tinzaparin under the name of Innohep. The fact that Novo Nordisk never satisfactorily managed to solve the problem of getting access to raw heparin for the production of Tinzaparin, most probably added to the decision to sell it off.

The attitude to the Tinzaparin project at Novo Nordisk was extremely ambivalent and, early during the development, a few myths about the project were established such as (1) "LMW heparin does not work," (2) "we will be number 7 or 10 in the market," (3) "heparin is a commodity drug" meaning that it would earn no money for the company," and (4) "what are you really doing? This project has been ongoing for ever." In fact, it had taken 9 years to develop and a normal time for developing new medicine is regarded to be 10 years! This experience was, to me, another example of how difficult it is for a company to enter into a new therapeutic area.

3.2.4 Substances to dissolve clots (thrombi)

Short after my start at Novo Nordisk in 1983, I became responsible for the department working on heparin and the thrombolytic substance made from pig plasma (Lysofibrin). The hemostasis research department was developed out of this group. We had several skilled biochemists and laboratory engineers who later became the core of the further development of hemostatic products at Novo Nordisk. However, it was not a well-functioning department, and my task was to widen the area of research and to improve cooperation in the department.

Regarding a widened range of hemostasis products, I thought, first of all, that the development of Tinzaparin should be the product to follow standard heparin and the first product in our hemostasis portfolio. The next step would be to include a substance to dissolve existent clots, preferably a tPA-product (tissue plasminogen activator). This would then be a follow-up of Lysofibrin. These products would cover the area of thrombosis therapy among the hemostasis products.

Collaboration had already been established with ZymoGenetics, Seattle, USA, with special focus on developing the knowledge on gene technology in mammalian cells. As Novo's main interest was in the area of thrombosis (heparin and pig-plasmin), the aim of the first project was to develop a fibrinolytic product that could dissolve blood clots that had formed in both arteries and veins. A plasminogen activator, tPA, was of special interest in the treatment of acute myocardial infarction. At this time, development of tPA preparations was very popular and several pharmaceutical companies were already well established in this area. GenenTech already had a patent on the physiological tPA molecule, which made it important to find a modified molecule, which should preferably have some special advantage over the normal molecule and above all be free from the problem of a patent. Such a thrombolysis product would also be a modern follower of Lysofibrin, the pig-plasmin.

3.2.5 The introduction of FVII at Novo Industri

Regarding the second part of the product portfolio in the area of hemostasis, products for stopping bleeding, I started to consider that it might be possible to dust off my idea of developing the recombinant factor VIIa (rFVIIa). Classical bleeding disease, hemophilia, is by far the most serious bleeding disease we know and thus the ultimate test for a hemostatically active product. My initial tests of the pd-FVIIa, which, in 1981 and 1982, had succeeded in stopping hemorrhages in two patients with severe hemophilia, complicated by inhibitors, clearly indicated that rFVIIa could be an important product for our hemostasis portfolio. However, I needed to test pure FVIIa on more hemophilia patients before I could recommend a complete gene-technological project for the production of rFVIIa for treatment of hemophilia patients.

My group at Novo had the necessary competence to produce pure products from plasma to treat patients. Both Lysofibrin and Tinzaparin/standard heparin were manufactured there. As I remember, I had several discussions with the Head of Research, Ulrik Lassen, about this idea and presented the results that Walter Kisiel and I had showed with the pd-FVIIa in the summer of 1983. I also made the suggestion that human plasma should be bought from the Finnish Red Cross, who were known for the high quality of their plasma (careful and extensive testing of blood donors and plasma regarding bloodborne pathogens such as different sorts of hepatitis viruses and HIV), for the production of a limited amount of FVII using the purification process that Walter Kisiel and myself had used [20].

At this point in time, Novo Nordisk had already started collaborating with ZymoGenetics to develop recombinant insulin and recombinant tPA for the dissolution of blood clots. As a step in a desire to further develop the hemostasis area, there were intensive discussions during 1983 and 1984 as to which coagulation factor would be appropriate for collaborating with ZymoGenetics from the point of the competence existing at Novo Nordisk. I was thus given the task of compiling several documents summarizing what was known about the treatment of hemophilia and different coagulation factors with a potential to improve this treatment.

In this connection, I also wrote, in May 1984, a plan for a limited project to confirm the hemostatic effect of FVIIa in hemophilia patients with and without antibodies. To my delight, the project was approved.

This was the start of the later project for developing recombinant FVIIa (rFVIIa). It meant a lot of work in obtaining Finnish plasma, presenting the project for my research group and going through the different parts with them. Work with producing pure pd-FVIIa from the Finnish plasma proceeded quickly, thanks to the competence of the group at Novo Nordisk. With permission from the relevant Ethical Committees, patients were recruited through colleagues in Gothenburg, Sweden (Dr. Lilian Tengborn) and Aarhus, Denmark (Dr. Stener Stenbjerg). Most of the patients treated with pdFVIIa had knee-joint bleedings, and positive results were reported in most cases [21]. In all, we treated five hemophilia A patients with inhibitors in seven bleeding episodes. One patient with hemophilia B without inhibitors was treated in conjunction with the extraction of a wisdom tooth. I had already come to the conclusion at this time that a high initial dose seemed to give the best effect by immediately initiating an effective local hemostasis. It also seemed to reduce the need for repeated doses. The pd-FVIIa was also given to seven patients with a low platelet count.

In 1984–85, I had shown that pd-FVIIa shortened the coagulation time in rabbits with a low platelet count (after the administration of platelet antibodies) [22]. This first work with pd-FVIIa in the treatment of bleeding was very rewarding, and all the biochemists, process engineers, etc., in the group were enthusiastic for the project. Of course, I described the disease hemophilia for them, its characteristics, available treatment, and why hemophilia patients, especially those with inhibitors needed improved treatment options. My 10 years of practical experience as a hemophilia specialist at the Hemophilia Unit in Malmö was of course a great help. You "live with" hemophilia patients and get to know them and their families extremely well. I shared some of this with the group, which, from being rather divided, grew into a tightly welded group of scientists who became more and more committed in their efforts to find better treatment options for these patients.

REFERENCES

[1] Kakkar VV. An international multicenter trial. Prevention of fatal postoperative pulmonary embolism by low doses of heparin. The Lancet 1975;2:45–51.

[2] Andersson L-O, Barrowcliffe TW, Holmer E, Johnson EA, Söderström G. Molecular weight dependency of the heparin potentiated inhibition of thrombin and activated factor X effect of heparin neutralization in plasma. Thromb Res 1979;15:531−41.

[3] Kakkar VV, Murray WJG. Efficacy and safety of low molecular weight heparin (CY216) in preventing postoperative venous thrombo-embolism: a cooperative study. Br J Surg 1985;72:786−91.

[4] Bergqvist D, Hedner U, Sjörin E, Holmer E. Anticoagulant effect of two types of low molecular weight heparin administered subcutaneously. Thromb Res 1983;32:35−44.

[5] Mätzsch T, Bergqvist D, Hedner U, Østergaard P. Effects of an enzymatically depolymerized low molecular weight heparin as compared with conventional heparin in healthy volunteers. Thromb Haemost 1987;57:97−101.

[6] McLean J. The thromboplastic action of cephalin. Am J Physiol 1916;41:250−7.

[7] McLean J. The discovery of heparin. Circulation 1959;19:75−8.

[8] Howell WH, Holt E. Two new factors in blood coagulation—heparin and pro-antithrombin. Am J Physiol 1918;47:328−41.

[9] Howell WH. The purification of heparin and its chemical and physiological reactions. Bull Johns Hopkins Hosp 1928;42:199−206.

[10] Jorpes E. The chemistry of heparin. Biochem J 1935;29:1817−30.

[11] Crafoord C. Preliminary report on postoperative treatment with heparin as a preventive of thrombosis. Acta Chir Scand 1937;79:407−26.

[12] Crafoord C, Jorpes E. Heparin as a prophylactic against thrombosis. J Am Med Assoc 1941;116:2831−5.

[13] Crafoord C, Norberg B, Senning Å. Clinical studies in extracorporeal circulation with a heart-lung machine. Acta Chir Scand 1957;112:219−45.

[14] Radegran K. The early history of cardiac surgery in Stockholm. J Card Surg 2003;18:564−72.

[15] Laurent TC. Studies on fractionated heparin. Archive Biochem Biophys 1961;92:224−31.

[16] Brinkhous KM, Smith HW, Warner ED, Seegers WH. The inhibition of blood clotting: an unidentified substance which acts in conjunction with heparin to prevent the conversion of prothrombin to thrombin. Am J Physiol 1939;125:683−7.

[17] Abildgaard U. Highly purified antithrombin III with heparin cofactor activity prepared by disc electrophoresis. Scand J Clin Investig 1968;21:89−91.

[18] Hull RD, Raskob GE, Pineo GF, Rosenbloom D, Evans W, Mallory T, et al. A comparison of subcutaneous low-molecular-weight heparin with warfarin sodium for prophylaxis against deep-vein thrombosis after hip or knee implantation. N Engl J Med 1993;329:1370−6.

[19] Hull RD, Raskob GE, Pineo GF, Green D, Trowbridge AA, Elliott CG, et al. Subcutaneous low-molecular-weight heparin compared with continuous intravenous heparin in the treatment of proximal vein thrombosis. N Engl J Med 1992;326:975−82.

[20] Hedner U, Kisiel W. Use of human factor VIIa in the treatment of two hemophilia A patients with high-titer inhibitors. J Clin Investig 1983;71:1836−41.

[21] Hedner U, Bjoern S, Bernvil SS, Tengborn L, Stigendahl L. Clinical experience with human plasma-derived factor VIIa in patients with hemophilia A and high tier inhibitors. Haemostasis 1989;19:335−43.

[22] Hedner U, Bergqvist D, Ljungberg J, Nilsson B. Hemostatic effect of factor VIIa in thrombocytopenic rats. Blood 1985;66(Suppl 1):289a.

CHAPTER FOUR

The Development of Recombinant FVIIa (rFVIIa) (1985–88)

Contents

4.1 BACKGROUND AND START

My first studies with plasma-derived FVIIa (pd-FVIIa) and the treatment of two hemophilia patients were carried out in Malmö at the beginning of the 1980s (see Chapter 2: My Encounter With Hemophilia (1959–82)). Already in Seattle during 1978–80, my collaboration with Walter Kisiel in Earl Davie's laboratory at the Department of Biochemistry, University of Washington, was established and my idea of using FVIIa in the treatment of hemophilia patients, especially those with inhibitors, was discussed in the laboratory. Thus, the Seattle group was well aware of my interest in developing better treatment for these patients at the end of the 1970s. In their work on developing techniques to transfer capacity to produce coagulation factors into mammalian cells, FVII was given a place of priority.

As a result, a hybrid genome construction, consisting of a part of the FVII gene and a part of the FIX gene, was introduced into a mammalian cell line (baby hamster kidney (BHK) cell) by the Seattle group in close collaboration with the molecular biologists at ZymoGenetics. This cell line then produced human FVII consisting of one part of FIX and one part of FVII already at the beginning of the 1980s. On June 1, 1984, Novo signed a contract with ZymoGenetics to develop techniques for the production of rFVII, and in January 1985, a project was started with the aim of constructing the complete DNA sequence of human FVII. Around Christmas 1985, this was sent over to ZymoGenetics and in January 1986, the final BHK cell line with the complete FVII gene could be sent back to the research group at Novo Nordisk. The FVII produced from this BHK cell proved to have the structures necessary for the molecule to function normally in the coagulation system.

The period 1984—86 was, thus, characterized by close collaboration between the research groups at Novo Nordisk and ZymoGenetics. The scientists at ZymoGenetics had learned a great deal about mammalian cells, and recombinant technique through the tPA project already started together with Novo Nordisk.

4.2 WORK ON RECOMBINANT FVIIa

In June 1985, Novo's upper management approved a project to develop rFVIIa for the treatment of hemophilia patients with inhibitors. This decision had been preceded by a period during which Novo's management was continuously informed of the results of the individual patients who were being treated with pd-FVII. During this period, I was summoned to several different types of management groups to present the idea that better treatment of hemophilia patients with antibodies could be achieved with the help of rFVIIa. On these occasions I concentrated on describing the patients and their situation, of which I had much experience from my work at the hemophilia clinic in Malmö. Of course, I also described in detail the effects that the addition of FVIIa had produced in earlier tests and studies.

At this time, I clearly sensed that Novo's chief executive officer (CEO), Mads Øvlisen, understood the need of better therapy for these patients. Providing that rFVIIa lived up to expectations, it would fit in with Mads Øvlisen's vision for Novo Nordisk, namely to produce drugs that would really "make a change" for patients. I was impressed that, without any medical training, he quickly grasped the idea of using rFVIIa for the treatment of hemophilia patients with inhibitors.

My impression was also that Ulrik Lassen, the head of research, and Knud Aunstrup, the head of the enzyme research, were attracted by the challenge posed by gene technology using mammalian cells. At this time, this technique had not been used to any great extent, and it was clear that production of complicated molecules like FVII and other coagulation proteins required mammalian cells. Neither yeast cells, used for the production of insulin, nor coli bacteria, used for the production of growth hormone could be utilized.

However, it was not easy to clearly deduce Novo's attitude to such an adventurous project. What was clear was that the financial and marketing sections were doubtful, given the limited number of patients who would benefit from the medicine. I had several times mentioned the idea that FVIIa would also be able to stop other hemorrhages and serve as a general hemostatic agent. However, apart from the few patients with low number of platelets (thrombocytopenic patients) who had been treated with pd-FVIIa (see Chapter 3: The First Years at Novo Nordisk), I had no further evidence of efficacy than the experiments on thrombocytopenic rabbits in Malmö.

In one of my documents, dated December 22, 1983, I mentioned that if "FVIIa turns out to be as efficient as the preliminary data suggest, also treatment of traumatic bleeding episodes and emergency should be added to the yearly requirement estimated for hemophilia patients." In the same document, I also pointed out that FVIIa could possibly also be used to treat patients with von Willebrand's disease. The document in question had been written at the request of the research management at Novo Nordisk. I shall later return to the possibility of using FVIIa even for other bleeding episodes than those of hemophilia patients.

It was, thus, with great joy and inspiration that I received Novo's decision on June 30, 1985, to approve a project to develop rFVIIa for the treatment for hemophilia patients with inhibitors.

4.2.1 The organization of the hemostasis group

A working group focusing on the development of rFVIIa was formed on the lines of a model previously employed by Novo called "the satellite model." It was made up of people with specialized knowledge from several functional areas loosely brought together to form a group. The composition of the group could be altered with changing needs. My experience of this organization was extremely positive.

Specialists in molecular biology with cell biology and other specialists in mapping out the protein structure such as the amino acid sequence and carbohydrate content were included. The rFVIIa should preferably be identical with the FVIIa in human plasma. We had also a pharmacist with the knowledge of how to stabilize pure protein and formalizing it to make it suitable for intravenous injection. An exceptionally important person in our group was an engineer specialized in process chemistry, Kurt Pingel. He had experience of the purification processes suitable for production in an industrial scale. From the beginning, it was important to avoid processes that could never be used in reality when producing rFVIIa for clinical purposes. Walter Kisiel acted as a scientific consultant to the group. I eventually succeeded in creating a group including pharmaceutical, assay technique, immunology, protein chemistry, and large-scale production expertise.

Although still very small, we were a highly dedicated group prepared to solve all kinds of problems. The development of rFVIIa was actually the first time a protein requiring mammalian cells for posttranslational modifications was produced in large scale. An extremely important part was the close cooperation with the scientists of Novo's enzyme division, NovoZymes, which for decades had produced industrial enzymes by using living organisms in fermentation processes. Representatives from this division were included in our group and were, in my opinion, enormously important with their knowledge of how to make living organisms, in this case BHK cells, thrive, grow, and produce human FVII in the big steel tanks. At this time, such a technique was still in its cradle, and Novo Nordisk had no experience of the fermentation of mammalian cells. Thanks to the expertise of the enzyme division's research department, the fermentation problem was solved surprisingly quickly.

I still feel great admiration for the competence that made it possible to produce sufficient rFVIIa to treat the first patient in conjunction with a surgical procedure in 1988, 3 years after the start of the project. In this short time, the problems with optimal stirring speed and suitable substrate composition to help the cells thrive were solved, allowing an increase in production volume to 500 L and later to 2000 L in 1987.

4.2.2 Production challenges

A problem that had to be solved was that the BHK cells produced the nonactive form of FVII, the so-called proenzyme. To be functional, the molecule had to be split between two defined amino acids (numbers 152 and 153). The activated protein, which then consists of two amino acid chains linked together with the help of a special binding, should form a complex with tissue factor exposed by a vascular damage. The final functional product rFVII thus must be activated to rFVIIa.

We worked intensively in our group to find a way of activating rFVII, preferably without adding new enzymes. As one of the important reasons for using a recombinant technique to produce rFVIIa for treating patients was to avoid the transfer of bloodborne infections such as hepatitis and HIV, we were anxious to avoid additives from human or animal sources. The solution to this problem came quite unexpectedly when it was observed that activated FVII molecules were spontaneously formed during the purifying process, when the FVII protein was adsorbed to positively charged substances. As the concentration of FVII protein increased during the adsorption process, rFVII was spontaneously changed into rFVIIa. The surprising observation was immediately published by the protein chemists at Novo [1]. This was a significant step in process development as it made it possible to exclude the addition of all other substances during manufacture. Thus, the final product consisted solely of pure rFVIIa.

4.2.3 The development of necessary assays

In the beginning, another great problem was the development of a stable and reliable method for determining factor VII activity. In this connection, we started collaborating with Trevor Barrowcliffe at the National Institute of Biological Standards and Control (NIBSC) in Great Britain. Trevor Barrowcliffe was, at this time, the head of the section for standardization of hemostasis products at NIBSC. Together, we carried out a study of our method of measuring the function of FVII in a coagulation system. In this study, samples with a fixed content of FVII were sent to several laboratories where they were analyzed with the method used at the laboratory in question. The obtained results were then compared. This collaboration led to Novo's rFVIIa being accepted by NIBSC as the International Standard for FVII determination.

However, the work of developing rFVIIa demanded much more work on development of new analyses, and, fortunately, we had several scientists in our group who were experienced in method development. They came from the original hemostasis group at Novo with experience from the development of heparin and low molecular weight heparin. They had also worked with pig-plasmin and were active in the recently started project to produce recombinant tissue plasminogen activator. This group eventually developed into an analysis group within the hemostasis research group and was given the responsibility for ensuring that all the rules for method standardization necessary for licensing were meticulously followed.

The importance of the specialized knowledge that this group developed for the continued production of rFVIIa and the further development of hemostatic products at Novo Nordisk cannot be overemphasized. In my opinion, it was an unwise decision to later break it up and incorporate its members in the larger section for method development where their competence was not appreciated. It also had the result that several of those who had substantially contributed to the successful development of rFVIIa left the company at the end of the 1990s and beginning of the 2000s.

One very special area, demanding great attention, was the development of analyses to determine potential impurities in the rFVIIa preparation. It was essential to find out which parts of the production could leave behind traces.

4.2.4 Characterization of the rFVIIa molecule

It was important for us to characterize our rFVIIa regarding amino acid structure, carbohydrate groups and their location, and gammacarboxylation, i.e., the modified form of an amino acid with extra carboxylic acid groups, which is necessary for the factor VIIa protein to be active. Already from the beginning, it was essential to me that our recombinant protein was as identical as possible to normal plasma FVII to minimize the risk of antibody development. We were happy to demonstrate that the rFVIIa had the same amino acid sequence as plasma FVII that the carbohydrate groups were in the same location and that they had virtually the same structure as found in plasma FVII. A comprehensive characterization of rFVIIa was published in 1988 [2] (Fig. 4.1).

4.2.5 Effects and side effects: preclinical animal studies

During the process of development, we came across new problems everyday. No one had previously done anything similar and we had no "guidelines" to follow. Our decisions were made based on common sense

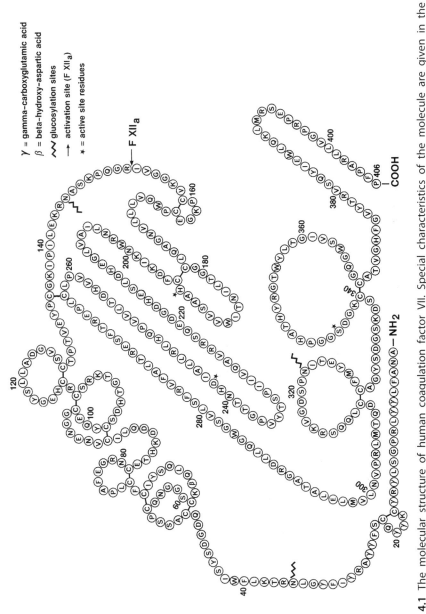

Figure 4.1 The molecular structure of human coagulation factor VII. Special characteristics of the molecule are given in the figure. Modified from E.W. Davie, K. Fujikawa, and R. Woodbury.

and our current knowledge. One of the advantages of the satellite model was the immediate feedback available from the other specialist fields. The group was soon joined by a pharmacologist, a veterinarian, who was given responsibility for the animal studies necessary to demonstrate that high doses of rFVIIa did not produce any side effects, and above all, that it did not induce thrombosis.

It was not entirely easy to convince the management at Novo to include a pharmacologist specialized in hemostasis in the research group. There was a tradition at Novo to distinguish between activities concerning chemistry and those involving testing in animals. However, after several visits to the directors responsible for product development and pharmacology and finally to the head of research, I succeeded in persuading them to place a pharmacologist in our group "temporarily, on trial."

This was a tremendous breakthrough for us and enabled us to develop animal models, the first and foremost to study safety regarding possible side effects but also models that would give us an idea of the ability of rFVIIa to stop bleeding in situations other than hemophilia. "Our pharmacologist," Viggo Diness, soon realized the advantage of being surrounded by colleagues with specialist knowledge in bleeding and thrombosis. With my 10 years of clinical experience in treating patients with these problems, I could also contribute with ideas on how the models could best mirror clinical reality and thus help us to find the best possible answers to our questions.

We embarked on this work with great energy and enthusiasm. We had been fortunate in getting an enthusiastic and exceptionally meticulous pharmacologist, who later contributed to the success of several other projects at Novo Nordisk. Apart from ours, I have never seen any projects with "FVIII by-pass therapy," which have so carefully studied the ability of their preparation to enhance the septic effect of the so-called endotoxins in rabbits (rabbits are especially sensitive to this). With the model worked out in our group, we could show, at a very early stage, that even high doses of rFVIIa had no septic enhancing effect [3]. This was important, as patients with severe hemorrhages often have complications in the form of serious infections and sepsis.

It was also important for us to prove that rFVIIa did not give rise to a general activation of the coagulation system. As early as the 1970s, I had demonstrated that activated prothrombin complex concentrate (aPCC) induced such changes in dogs [4]. We associated them with the serious side effects that had been reported in the use of aPCC [5,6]. It was therefore imperative to demonstrate, in different rat and dog models, that rFVIIa did not have such side effects. At the same time, we could establish that aPCC (FEIBA), contrary to rFVIIa, initiated changes in the coagulation factors indicating a systemic activation of the coagulation system [3].

This was, in my opinion, extremely important for the entire development of rFVIIa. Our ultimate aim was, of course, to produce a preparation that had no side effects. The studies we carried out would have been impossible with the exceptionally limited amounts of pd-FVIIa, which Walter Kisiel and I had produced in 1980.

The hemostatic effect of rFVIIa was finally tested on hemophilia dogs at Chapel Hill, NC, USA. The colony of dogs with hemophilia was established in Chapel Hill at the end of the 1940s by Dr. Kenneth Brinkhous, who was Professor of Pathology of the University of North Carolina at Chapel Hill, and developed into a treatment center for dogs with hemophilia. For many years, this was the only colony of hemophilia dogs in the United States, and the majority of hemophilia products have been tested in these dogs. Dr. Brinkhous took great care of his dogs, and they were looked after impeccably. Under the auspices of Dr. Brinkhous and later Dr. Harold Roberts, Chapel Hill developed into one of the world's best hemophilia centers. Clinical work with hemophilia at this center (Hemophilia Comprehensive Care Center) was combined with hemostasis research of the highest quality.

The research group at the University of North Carolina at Chapel Hill had, for many years, received continuous financial grants from the National Institute of Health (NIH). I had known Dr. Roberts for a long time through the International Society of Thrombosis and Hemostasis (ISTH), where he was the secretary for many years, but I had never visited Chapel Hill. We had even discussed coagulation research in different connections since the 1970s when I had participated in various international research groups and meetings. However, my first visit to Chapel Hill took place in January 1988. The main objective of my visit was to discuss with Dr. Brinkhous the possibility of testing rFVIIa in his hemophilia dogs. He and his research group had developed a model for testing the hemostatic effect of various agents in hemophilia [7].

In this model, the bleeding following a standardized nail clipping was measured before and after the administration of a potential hemostatic agent. Dr. Brinkhous was, at first, very hesitant as to whether he should let me test rFVIIa in his dogs. He did not feel confident about the effect of rFVIIa and did not want his dogs to be submitted to any risk whatsoever. After I had described the theory behind the idea of rFVIIa and showed him all the facts we had about its effect on both animals and humans, he finally reluctantly agreed, and it was decided that I would return later to carry out the study. At that time Dr. Brinkhous personally led the experiment. He designed the exact study protocol with me and was present during the tests. Bleeding time was measured before and after injection with rFVIIa. The effect of rFVIIa was clear. Bleeding time was normalized and Dr. Brinkhous became more and more enthusiastic. Together, we wrote an article that was accepted for oral presentation (Dr. Brinkhous) at that year's meeting of the American Society of Hematology in New Orleans [8].

4.3 TREATMENT OF THE FIRST PATIENT WITH rFVIIa, 1988

In January 1988, before we were ready with the IND document (Investigation of a new drug), I was contacted by a colleague, Hans Johnsson, at the Karolinska University Hospital in Stockholm. He had a patient with severe hemophilia with inhibitors who badly needed knee joint surgery on account of hemoarthrosis accompanied by acute pain. His need for painkillers was accelerating and to break this vicious circle required an operative procedure.

My colleague's question was whether we had any pd-FVIIa that could be used in this patient. However, nothing was left over from our previous studies to confirm the effect of FVIIa on hemophilia bleeding, and I had to tell him that we did not intend to produce any more. I did, however, tell him that we had rFVIIa, but that we were not quite ready with the document for the IND file.

My colleague then contacted the Swedish Health Authorities to inquire about the possibility of obtaining a license to use our rFVIIa preparation to be able to perform the necessary knee surgery. He spoke to the same person at the Health Authority Office, whom I had been in contact with in connection with the development of pd-FVIIa several years earlier, at the hospital in Malmö. She was, therefore, well acquainted with the background for using FVIIa to prevent or stop hemorrhages in hemophilia patients and she, herself, also pointed out that using a recombinant preparation should be superior to a plasma-derived one, bearing in mind the risk of transmitting various virus diseases. She suggested, therefore, that Novo Nordisk should send over the documents that were ready so that she could decide whether a license could be granted for the patient in Stockholm. This we did, and a license to use rFVIIa in connection with the planned knee joint surgery was issued.

In this connection, I would like to emphasize the importance of my colleague's initiative in expediting the development of rFVIIa. Because of it, we got proof that rFVIIa worked as bleeding prophylaxis during an orthopedic surgical procedure in a patient with severe hemophilia at least 6 months earlier than if we had been forced to wait for the IND permission to start clinical studies.

A period of careful planning preceded the operation that was to take place on March 9, 1988. One of the great problems with the introduction of rFVIIa was, and still is, to find the right dosage. The use of rFVIIa as a hemostatic product is a completely new concept in hemophilia treatment. It was not clear how much extra FVIIa would be needed to initiate hemostasis.

The dosage I estimated for the patient who was to undergo knee surgery on March 9, 1988, was largely based on the results of test tube experiments and, to some degree, on animal studies. A careful plan was worked out for the repeated injections of rFVIIa starting just before intubation required for continued anesthesia. In this planning, I made use of the treatment plan, worked out in detail by Inga Marie Nilsson for use in hemophilia surgery and which I had learned during my time at the Coagulation Clinic in Malmö. One of the main points in this treatment plan also was the importance of administering the concentrate to be used immediately before the first challenge on hemostasis, in other words, just before the introduction of the pharyngeal tube. The continued management of hemophilia surgery, according to the Malmö model, includes a regular supply of the hemostatic concentrate every 2—3 hours during the first 24 hours and, not least important, close observation of the patient during and after the procedure. No wounds should be closed before complete hemostasis has been established. Treatment is also given to protect the fibrin plugs that are formed from dissolving too soon. To achieve this, antifibrinolytic therapy (tranexamic acid) is administered.

My training at the hemophilia clinic in Malmö had taught me that all these steps are crucial for a successful result when operating on hemophilia patients. In this, there is no difference between hemophilia patients with or without inhibitors. However, the presence of antibodies constitutes an added risk as FVIII/FIX cannot be administered in cases of unexpected bleeding. Therefore, close observation is necessary. This was especially important in the first treatment with rFVIIa. Before rFVIIa became available, surgical procedures were advised against for hemophilia patients with inhibitors. It was, therefore, not surprising that I was nervous in face of what I regarded as the ultimate test of whether rFVIIa would live up to my hopes of replacing FVIII/FIX in hemophilia patients who had developed antibodies against FVIII or FIX. I had also discussed measures with my colleagues in Stockholm if it turned out that rFVIIa did not result in satisfactory hemostasis.

For our group in Copenhagen preparations included that a suitable number of rFVIIa ampoules were to be produced, marked, and packaged according to the rules for pharmaceutical products. The dry—frozen rFVIIa powder must be stored at $-80°C$, making the transport to Stockholm somewhat hazardous. As I did not trust the transport companies, I decided to take a box of ampoules with me to Malmö and place it in a $-80°C$ freezer at the hospital between Friday afternoon and Monday morning March 8, when I intended to fly from Malmö to Stockholm.

Early on Monday, I left home to catch the first plane to Stockholm, stopping on my way at the Department of MAS, where I had placed the valuable rFVIIa ampoules, to collect them. It was quickly done, and I fortunately did not discover that there had been a break-in during the night and that the burglar sat asleep in one of the offices.

The travel department at Novo Nordisk had booked me into a hotel at Solna Square, which is not exactly noted for its charm. The weather seemed to me typical for Stockholm in early March, windy and cold with uneven, icy road conditions made for causing accidents. It was altogether very inhospitable and unpleasant. I had forgotten to tell the travel department of Novo Nordisk that Solna Square is nowhere near the Karolinska University Hospital although the post address is Solna. On the contrary, it is difficult to get there from Solna Square. As it was important to get the box with the valuable rFVIIa ampoules to a freezer as quickly as possible, I set off across the slippery Solna Square to catch a bus that would take me to the hospital.

To my dismay, when I finally arrived at the hospital and the Coagulation Laboratory there, I was met by cries of "Oh, was it today you were supposed to come." I could not believe my ears and said that I assumed that everything was meticulously prepared. I wanted to speak to the doctor who was to perform the surgery the next day but was given the reply that they did not know who it would be. Unnecessary to say, I felt deeply unhappy and tried, without offending anyone, to say that under no circumstances I was leaving the coagulation laboratory before I had spoken to the surgeon.

Thus, I spent most of the day sitting in the windowless, cramped "coffee room" of the coagulation laboratory. Finally, late afternoon, I was told who was going to operate and that I could meet him. It was proved to be a hand surgeon who was doing specialist training at the Orthopedic Clinic at the Karolinska University Hospital. His name was Akke Alberts (K.A. Alberts). For the first time that day I experienced infinite relief, bordering on happiness. A hand surgeon must understand the importance of complete hemostasis during operative procedures! It also turned out to be easy to explain to this pleasant, painstaking hand surgeon how important it was that he took "one layer at a time" and "did not proceed until he was certain that all capillaries were tight." To my great relief, he understood completely.

Relatively relieved, I returned to my drab hotel where I had a solitary dinner (there were no other guests in the dining room, which I could well understand). The night was not entirely restful and at about 5 o'clock in the morning I woke up with a feeling of panic about what I was about to do. If it had not been so unsuitable to start calling people at 5 o'clock in the morning, I may very well have phoned to cancel the whole thing. But it did not seem to be the best thing to do and, instead, I went through all the data I had in my head, which spoke in favor of rFVIIa acting as a hemostaticum and most likely also during surgery, which in itself is the most sensitive test of a substance's ability to prevent hemorrhaging. It is impossible to perform surgery in hemophilia patients without severe hemorrhaging if they are not treated with an effective hemostatic agent. The result of this thorough internal review was that I once more came to the conclusion that "rFVIIa *must* work." Furthermore, I would never know if it worked or not, if I did not embark on this venture and saw it through to the end! After this intellectual inquisition, I slept for a while before taking myself on shaky legs to the hospital in the morning of March 9, 1988.

My colleague from Copenhagen, Steven Glazer, who was responsible for the clinical development of rFVIIa joined me that morning in the operating theater. It was, of course, my task to dissolve the number of ampoules calculated to be a suitable dosage and to ensure that it was injected immediately before intubation. Steven Glaser contributed to the dose being somewhat higher than I from the beginning had suggested by saying "finish the ampoule" when it turned out that my suggested dose would leave about 1 ml in the ampoule. I am deeply in debt to him for those words as it became obvious during the later development of rFVIIa that the dosage, in general, should be increased to obtain the best effect.

When the first dose of rFVIIa had been injected, and the patient anesthetized and intubated, the surgeon started with a skin incision and found that the capillary bleeding stopped, whereby he could cut through the following layers of muscle and finally open the knee joint. In conjunction with this the blood that had accumulated in the patient's knee joint was suctioned out, accompanied by a loud slurping sound. It always does so, and when the surgeon saw me grow pale and sit down on a stool, lowering my head not to faint, he said comfortingly, "There is *always* blood in an inflamed joint." I knew this, of course, but I shall never forget how awful that slurping blood suction sounded imitating an ongoing bleeding.

This was, however, the only blood in the suction apparatus during the entire procedure. Instead, when I leaned over the surgical wound, I could see how strings of fibrin were formed in the wound, which was proof that rFVIIa worked and did, indeed, lead to the formation of fibrin plugs in the wound. It was a wonderful feeling! The continued dosage of rFVIIa during the day after surgery was administered by my colleague, Sam Schulman. He was fantastic in keeping track of all the planned blood sampling and dosages, which is so important in all hemophilia surgery. In this work he used a study protocol for the clinical testing of Novo Nordisk's low-molecular heparin, Logiparin (Tinzaparin), which was in progress. We had no study protocol ready for use of rFVIIa in hemophilia patients. It was, however, important that all data be recorded for future reports, so I had brought the Logiparin protocol with me to Stockholm. I went carefully through it with him. He also saw to it that the samples were immediately centrifuged and that the plasma was then separated from the red blood cells and placed in a suitable freezer box. These steps are important to ensure reliable test results.

An interesting incident occurred when the tests were being analyzed at the Central Laboratory of the Karolinska University Hospital. I received a telephone call from the person there, responsible for carrying out the coagulation analyzes. He was very upset and pointed out to me that there must be something wrong with the entire sampling procedure and treatment as the special test used for the determination of the so-called fibrin monomers, which is used as a sign of a generalized activation of the coagulation system, was not positive despite that the patient had been given rFVIIa. He pointed out indignantly that FVIIa was known to give intravascular coagulation and that there *must* be soluble fibrin monomers in circulating blood after injection of FVIIa. I tried in vain to explain to him that this was the very point with FVIIa that it did *not* give rise to circulating thrombin and fibrin.

When the operation was over and the patient was taken to the recovery room before going onto the ward, Steven Glaser and I had lunch at Stallmästaregården a well-known restaurant close to the Karolinska University Hospital. He was going back to Copenhagen, while I stayed until the next day to check on the patient. The next morning I observed that there was no blood at all in the drain placed in the operation wound, which indicated an effective hemostasis. The drain was removed, and I went back to Copenhagen very tired but extremely relieved and happy.

The next day when I came to Novo in Bagsvaerd, where I had my office, I happened to meet Mads Øvlisen in the reception and was given a big hug and congratulations. He had already heard about the successful operation. I had immediately after the operation phoned to my coworkers in the hemostasis group in Copenhagen to tell them about the successful result. After my phone call, the news had immediately spread to large parts of Novo and also reached the CEO, Mads Øvlisen, who shared my joy and satisfaction.

At this stage in the development of rFVIIa at Novo, it was necessary to carefully go through the resources in terms of personnel and finances required for the toxicological studies necessary before the start of the clinical studies. When I first started the project at Novo, I thought that the decision to invest in the development of hemostatic products and the necessary research, more or less automatically implied that there would be sufficient financing to carry out both the preclinical and the clinical studies and to develop the necessary production technique. This, however, did not appear to be the case. Every request for increased resources met with insurmountable obstacles. In August 1988, almost 6 months after the successful operation, the project leader for recombinant hemostatic products including rFVIIa, Kurt Pingel, drew up a plan calculating the necessary resources. The operation in March had generated pressure from other clinics where complicated hemophilia patients with inhibitors were not able to get adequate treatment. In Kurt Pingel's plan, it was pointed out that existing resources barely served to cover current production needs. There were no resources for production improvements, necessary preclinical studies or the development of methods.

Unfortunately, most of the continued development of rFVIIa was characterized by a constant struggle for necessary resources, which clearly slowed development. I will come back to this constant problem later.

REFERENCES

[1] Bjoern S, Thim L. Activation of coagulation factor VII to VIIa. Res Discl 1986;269:564—5.
[2] Thim L, Bjoern S, Christensen M, Nicolaisen EM, Lund-Hansen T, Pedersen AH, et al. Amino acid sequence and posttranslational modifications of human factor VIIa from plasma and transfected baby hamster kidney cells. Biochemistry 1988;27:7785—93.
[3] Diness V. rFVIIa in an endotoxin model and in a Wessler model. In: Hedner U, Roberts HR, editors. Proceedings of the second symposium on new aspects of hemophilia treatment. Asten, The Netherlands: Medicom Europe; 1991. p. 147—50.
[4] Hedner U, Nilsson IM, Bergentz SE. Studies on the thrombogenic activites in two prothrombin complex concentrates. Thromb Haemost 1979;42:1022—32.
[5] Kingdon HS, Lundblad RL, Veltkamp JJ, Aronson DL. Potentially thrombogenic materials in factor IX concentrates. Thromb Diath Haemorrh 1975;33:617—29.
[6] Pepper DS, Banhegyi D, Howie A, Cash JD. In vitro thrombogenicity tests of factor IX concentrates. Brit J Haematol 1977;36:573—83.
[7] Brinkhous KM, Sandberg H, Garris JB, Mattsson C, Palm M, Griggs T, et al. Purified human factor VIII procoagulant protein: comparative hemostatic response after infusions into hemophilic and von Willebrand disease dogs. Proc Natl Acad Sci 1985;82:8752—6.
[8] Brinkhous KM, Hedner U, Garris JB, Diness V, Read MS. Effect of recombinant factor VIIa on the hemostatic defect in dogs with hemophilia A, hemophilia B, and von Willebrand disease. Proc Natl Acad Sci 1989;86:1382—6.

CHAPTER FIVE

The Further Use and Development of rFVIIa (1989—96)

Contents

5.1 THE PROMPT USE OF rFVIIa

The week after the treatment of the first patient with rFVIIa in Stockholm, a meeting for hemophilia specialists was held in Spain. Only a small number of doctors are active in hemophilia care and they often meet in different places around the world. News in this area, therefore, travels fast. Sam Schulman, my coworker in Stockholm, attended the meeting in Spain and told his colleagues about the successful knee surgery, performed in a hemophilia A patient with inhibitors, under cover of rFVIIa. The news spread and I soon began to receive telephone calls from hemophilia centers around the world, inquiring about the possibility of obtaining rFVIIa for treatment of their patients with severe hemophilia with inhibitors.

The second patient was treated in May 1988 in Chapel Hill. Dr. Roberts, at the Hemophilia Center there, had a patient with severe laryngeal bleeding that was threatening to suffocate him. An unsuccessful attempt had already been made to treat him with an activated prothrombin complex preparation (FEIBA) [1]. Afterward, I received the following letter from Dr. Roberts: "There is no question that without treatment,

Treating Life-Threatening Bleedings.
DOI: http://dx.doi.org/10.1016/B978-0-12-812439-0.00005-8
53

the laryngeal hematoma in the patient would have resulted in death. Spontaneous resolution of this hematoma without treatment would constitute a major miracle requiring divine intervention" [2]. It brought great joy to all of us who had worked with the development of rFVIIa.

This was the start of a rapid expansion in the use of rFVIIa for treatment of complicated cases of hemophilia with inhibitors. Colleagues around the world asked if they could get hold of rFVIIa for treatment of their very complicated cases of hemophilia patients with inhibitors. It was obvious that rFVIIa could help these patients, but as the product was not yet licensed, we needed to get permission from the "Federal Drug Administration" (FDA) before each treatment.

However, after following the treatment of a number of hemophilia patients, the FDA decided that there was no risk of serious side effects and suggested that Novo started a so-called "Compassionate Use" study for these patients.

5.2 rFVIIa AS A "COMPASSIONATE USE" PRODUCT IN THE UNITED STATES

To make life-saving treatment available to special patient groups, the United States had, what they called, the "Compassionate Use Program." This program had been started to provide effective treatment for patients in serious situations, while the work to obtain full licensing of the drug in question was still underway. To be eligible for this program, the patient (in this case hemophilia patients with inhibitors) must be suffering from "life or limb-threatening bleedings" in which available treatment has failed.

The hemophilia patients included in this program were seriously ill with life-threatening complications. It was quite a challenge collaborating with the local doctor by phone and trying to start an effective treatment with rFVIIa. Steven Glaser drafted a protocol, according to FDA regulations, for these severely ill patients, and we sent rFVIIa to the different treatment centers after the required permission for treatment had been received at the site. There was a tremendous need, and we had soon treated a considerable number of patients in this category.

The result was impressive, and we could rejoice with the local treatment group in the fact that we had saved many patients' lives. In my first publication (1990) about my experiences of treating hemophilia patients with inhibitors with rFVIIa, I gave a detailed report of the first 20 patients successfully treated in our "Compassionate Use" program [3]. Among these patients, 13 had life-threatening hemorrhages and 9 were treated in connection with surgical procedures. Two of the life-threatening bleedings were cerebral hemorrhages. One patient, with acquired hemophilia, was treated in conjunction with a serious hemorrhage in the laryngeal region, and one had an FXI deficiency and successfully underwent a testicle operation.

A patient with hemophilia B was treated in connection with a severe forearm bleeding. To avoid permanent serious damage to the boy's nerves and circulation, major surgery had been performed in his forearm. During this procedure, there had been severe hemorrhaging and the patient risked losing his forearm if the bleeding could not be stopped. After a few doses of rFVIIa the bleeding ceased and the patient's arm healed without further problem. He was later given several skin grafts, under cover of rFVIIa [4]. The boy was very happy and showered Steven Glazer, who visited him in hospital during the acute phase, with praise and admiration for having found such a wonderful medicine. Steven Glazer tried to tell him that it was a female doctor behind the original idea and not him, but the boy absolutely refused to believe it. Later, this boy has appeared several times in different contexts and talked about the fantastic rFVIIa.

Unfortunately, a few patients could not be saved, for example, a patient with a massive abdominal hemorrhage that led to renal failure. This patient was treated early on in fall 1988. The hemorrhage was under control, but, unfortunately, there was not sufficient rFVIIa available at the time for the patient to be treated long enough. Despite the tragic end, the patient's parents wrote a letter to Steven Glazer in which they thanked him for Novo's efforts during their son's illness and especially for the fact that Novo had developed a drug for hemophilia patients with inhibitors. They expressed their profound gratitude and happiness that these patients would have access to better treatment in the future [5].

Thus, our "Compassionate Use" study included patients with cerebral hemorrhages (a life-threatening condition for hemophilia patients), severe abdominal bleeding, muscular bleeding in an arm, causing pressure damage to nerves and blood vessels, and several other life-threatening hemorrhages. Among them are many fantastic case stories. At this time Novo had no affiliate in the United States and the distribution of rFVIIa was dealt with by the Toronto office in Canada. In urgent cases the staff themselves transported rFVIIa to various hospitals in the United States. Thus, they became deeply involved in different cases and volunteered to help in the most amazing way. One of them delivered rFVIIa to the boy with the terrible forearm-bleeding and has kept in touch with him ever since [4].

Another one of the very complicated patients was a young man who had developed a large cyst (a so-called pseudo-tumor) in the abdomen around one of his hip joints. During an attempt to remove this extremely large cystic formation surgically, the patient developed a massive, diffuse hemorrhage in the operation site, which proved impossible to stop with the help of conventional therapy. However, after administering rFVIIa, the widespread bleeding was stopped. To complete the surgical procedure, the operation was resumed, and the patient once again began to bleed profusely, a hemorrhage believed to be caused by the activation of both the coagulation system and the clot lysis system. This can sometimes occur as the result of extensive tissue damage, as in this case. Unfortunately, the patient's life could not be saved [6,7]. This episode gave rise to a discussion among hemophilia specialists, as to whether rFVIIa could have contributed to the tragic result by inducing coagulation in small capillaries. In my opinion rFVIIa had, instead, helped to stop the severe, massive diffuse bleeding that most probably was caused by the extensive decomposition of tissue, resulting in the activation of a host of enzymes leading to extensive organ damages. Further experience with the use of rFVIIa has not shown any adverse effects of this kind.

In 1991, we had 52 patients and 112 were reported in 1993, who had been treated for severe bleeding or in connection with vital surgical procedures, according to our initial "Compassionate Use" protocols [7,8]. Thirty-seven of these had hemophilia A with inhibitors, four had Hemophilia B with inhibitors, eight had acquired hemophilia, one had a factor XI deficiency, and one had a factor VII deficiency. We also had one patient who did not have hemophilia but an acquired antibody that, after a blood test, was characterized as an "antithrombin-like" inhibitor.

Nine of these patients were treated in conjunction with vital surgical procedures. The first patient was the one who underwent surgery in Stockholm 1988, when rFVIIa was used for the first time. Patient 2 was operated on in Milan on 1989 for a very large inguinal hernia. It had not been possible to treat the repeated bleedings in his hernia sac effectively. The resulting sac was so big that the patient could not get out of bed.

The operation of a hernia like this involves exposing large areas of tissue with a considerable number of easily bleeding small blood vessels (capillaries). Thus, abundant bleeding was to be expected in an operation of this kind. Steven Glazer flew to Milan and was present during the operation. We had agreed on an initial dose of 75 µg/kg. The same dose was repeated after 2 hours and the third dose after 4 hours. A beginning oozing was then noted and a dose interval of 3 hours was introduced with the intention of continuing in this way for at least 24 hours. Steven Glazer reported on his return to Copenhagen that evening that the operation had gone well and no abnormal bleeding had occurred during the procedure.

However, 26 hours after the operation, bleeding from the incision was observed, which immediately ceased after an extra dose of rFVIIa. After a further 24 hours, a wound hematoma had developed. rFVIIa was then administered continuously by intravenous drop. The bleeding did not stop, and it was decided to switch to porcine factor VIII. Continuous infusion was not recommended by us, as we had previously noted that the FVIIa molecules were adsorbed on the walls of the plastic tubing, so that only minimal amounts reached the patient. We had therefore carefully instructed users of rFVIIa to administer it by intravenous injection directly into one of the patient's peripheral blood vessels. The hospital in Milan had also been given these instructions but had still used continuous infusion. They therefore drew the conclusion that rFVIIa did not work in conjunction with surgical procedures and published their views in an article with the title "Failure of recombinant factor FVIIa during surgery in a patient with high-titer factor VIII antibody" [9].

During a personal meeting later, with Professor Mannucci, head of the hemophilia center in Milan, he regretted for not having been in Milan when the patient was treated. He was quite clear that the treatment had not been carried out according to instructions. My conclusion from the experiences in Milan was that the dosage was not sufficiently high for use in hospitals without specialized experience of rFVIIa and hemophilia. In consequence, we increased the initial dose to 90 µg/kg but kept a dose interval of 3 hours.

Another two patients with large inguinal hernias were operated on under cover of rFVIIa. The first of these was in Helsinki, Finland, during Spring 1991. An initial dose of 90 µg/kg was used. The second dose was decreased to 70 µg/kg and given every hour. The same dose was then administered at 3-hour intervals. Signs of abdominal bleeding were observed the morning after the operation, and the dose was increased to 90 µg/kg every third hour. After a further 24 hours a new operation was carried out, removing an organized hematoma, which proved that the dosage of 90 µg/kg every third hour had effectively stopped the bleeding in question. The lesson I learned from this operation, where I was present during the whole procedure, was that a dose of 90 µg/kg was necessary to achieve complete hemostasis and that the dose should be repeated every third hour for at least 24 hours [7].

When the next patient with a large inguinal hernia came into question, I suggested that a surgeon, experienced in operating hemophilia patients, should be present to advise on the practical details that are so important for successful hemophilia surgery. The third operation was carried out in Australia in 1993. At this operation, both myself and a surgeon with more than 20 years' experience of surgery on hemophilia patients in Malmö were present. We used a dose of 90 µg/kg every third hour and a surgical technique with careful handling of the tissues, and absolute hemostasis, suturing all capillaries and ensuring a completely dry operating site. The patient was also given antifibrinolytic treatment (tranexamic acid), which was routine in Malmö when operating on hemophilia patients. This operation proceeded without any problems, and hemostasis was impeccable [10]. It was a great relief to me.

I could from these experiences establish that rFVIIa really was effective as a hemostaticum on the condition that the treatment was carried out with care, and that the dosage of 90 µg/kg was given every third hour for 24 hours.

The protocol we had used for treatment of patients with life-threatening hemorrhages in the so-called "Compassionate Use" program had, however, proved to be unnecessarily complicated and our growing experience of treating hemophilia patients with inhibitors with rFVIIa clearly indicated the need for a modified version, drafted by Steven Glazer in March 1991. At this time, there had been great changes at Novo, which had merged with Nordisk Gentofte to become Novo Nordisk A/S. A large-scale reorganization of most of the functions important for the development of rFVIIa was carried out in connection with this merger. I shall discuss about this later. A direct consequence was that the new protocol, ready for acceptance in March 1991, was never approved and the, much longed for, modification of the "Compassionate Use" treatment unfortunately was never carried out. This created substantial problems for the continued development of rFVIIa, during the 1990s.

5.3 PLANS FOR THE CLINICAL DEVELOPMENT OF rFVIIa

The first clinical development plan, drawn up by Steven Glazer, was presented in November 1987. The aim was to obtain a license for the use of rFVIIa in cases of acute hemorrhaging in hemophilia A and B patients, with or without inhibitors. The plan also included a study of rFVIIa in prevention of joint bleedings, the so-called prophylaxis. Another study aimed at studying the hemostatic effect of rFVIIa in surgery was included. Possibilities of using rFVIIa to stop or prevent bleeding in cases of von Willebrand's disease also affecting women, FVII deficiency, low platelet count, abdominal bleeding, and, finally, FXIII-deficiency were covered by further planned studies.

A revision of the first clinical development plan was presented in February 1988. It had a licensing strategy that aimed at quickly obtaining a license for the treatment of hemophilia A and B patients with inhibitors and at the same time investigating the possibilities of exploring the product's full potential. The plan consisted of a pharmacokinetic study (Study I) and a study of the effect on visible bleeding in hemophilia patients (Study II). Study III was designed, at this stage, to be a comparative study between rFVIIa and FEIBA (aPCC). According to the plan, these studies were to be carried out in two hemophilia centers in Great Britain and two in the United States. It was planned to be carried out as an analysis of bleeding episodes treated with either rFVIIa or FEIBA. The idea was to manufacture identical, unlabeled, bottles, containing each of the two substances. In this way, neither the patient nor the doctor would know whether the patient was being treated with rFVIIa or FEIBA.

Already in July 1987, Steven Glazer contacted Immuno AG in Vienna, the manufacturers of FEIBA, to find out if they were interested in collaborating with Novo in a comparative study of rFVIIa and FEIBA. After a reminder in October, Immuno AG replied that they wished to obtain rFVIIa for in vitro testing. Novo Nordisk could not meet their request, as rFVIIa was not available in ready-made ampoules at this time. However, on Novo's part, we continued to plan our comparative study of rFVII and FEIBA. During a personal visit to Immuno AG in Vienna, however, Steven Glazer and I were informed by Immuno's management that they were not interested in participating in any study involving recombinant drugs. The continued work of planning the study was also impeded, on Novo's part, by the increasing need to make rFVIIa available to seriously ill patients in the Compassionate Use program. Furthermore, the costs of such a complicated study would be too high to be borne by Novo Nordisk alone.

On November 1, 1988, an application for the "Investigation of a New Drug, IND" was sent in to FDA in the United States. It was approved in January 1989, and the clinical plan was essentially similar to the previous one from early 1988. A plan to continue the most efficient dose found in each patient from the controlled, randomized, double-blind study in a home treatment setting was, however, added. Unfortunately, this follow-up study was never carried out. The clinical development section of the Biopharmaceutical Division in charge of the development of rFVIIa at the time felt that resources for this follow-up could not be set aside. This was especially unfortunate because home treatment in mild to moderate hemophilia bleedings has been a generally accepted way of treating hemophilia patients since the early 1980s. It is well-known that treatment initiated within 2 hours of the first signs of a beginning hemorrhage is clearly more effective as compared to a treatment started later on [11]. Novo Nordisk also had to perform a home treatment study later on before licensing was approved in the United States.

Study I listed in the Clinical Plan of the IND focusing on effect and safety of rFVIIa in patients with or without ongoing bleeding was carried out on 1989—90, but the report was delayed until April 1991 due to the lack of resources in different functions of the Biopharmaceutical Division.

5.4 THE MERGER OF NOVO AND NORDISK GENTOFTE 1989 AND ITS CONSEQUENCES

The merger of Novo and Nordisk Gentofte in 1989 brought about a thorough reorganization into three divisions: (1) the insulin and diabetes division, (2) the division for the development of drugs acting on the central nervous system (CNS Division), and (3) the division for the development of bio-pharmacological drugs (Biopharmaceutical Division).

The research and development in the hemostasis area sorted under the last-named division. There was also a reorganization of the different function areas. Consequently, the satellite model that had worked so well in developing rFVIIa disappeared. Instead the hemostasis group was split up, and its members incorporated in the three functional groups—research, production, and development, including clinical development. At this time rFVIIa was regarded as a developmental project with no need of research support. It was only to be prepared for licensing, and research facilities were not allowed to be used for further investigation of rFVIIa.

However, it was obvious that the further development of rFVIIa would, indeed, need a lot of research support. Treating hemophilia with an addition of FVIIa was an entirely new concept and, for it to be approved as a new drug, I was convinced that the answers to various questions about its mechanism of action, etc. were needed. This required research resources. The new research section in the Biopharmaceutical Division was responsible not only for the development of new products within the hemostasis area but also for the further development of growth hormone, Nordisk Gentofte's most important biopharmaceutical drug. Therefore, enough resources for further studies of the mechanism of action of rFVIIa could not be provided.

As a consequence of these changes, I took the initiative to establish an intensive external collaboration with several research groups around the world. There was great international interest for rFVIIa from both clinicians and scientists. Therefore, many well-known research groups were willing to collaborate with us in solving different problems concerning the FVII-dependent hemostasis mechanism.

The Hemophilia Center and the group of Hemostasis and Thrombosis Research at the University of North Carolina at Chapel Hill headed by Professor Harold Roberts became our first coworker. I had known Dr. Roberts since the early 1970s from working with the Society of Thrombosis & Hemostasis and had, at an early stage, discussed the use of rFVIIa in the treatment of hemophilia with him. He had treated the second patient with inhibitors with rFVIIa, and his group participated in the first study of different doses of rFVIIa. It was also in Chapel Hill that I had carried out the first studies of the effect of rFVIIa on hemophilia dogs in Spring 1988.

The hemostasis process was, at this time, described according to the "waterfall" or "cascade" model from the 1960s, which was based on test tube experiments using artificial phospholipids as platelet substitutes. However, this model did not explain why patients with FXII deficiency do not bleed [12,13]. Also, the role of FVII/FVIIa was not well understood. It had been demonstrated that FVIIa lacked enzymatic activity, unless it was complexed with tissue factor (TF) [14,15]. A series of important observations including that the FVIIa-TF complex not only activates FX into FXa but also FIX into FIXa were summarized by Sam Rapaport and Vijay Rao [16,17]. The importance of the "tissue factor pathway" thus was stressed, and it was later concluded that the major initiating event in hemostasis is the formation of the FVIIa-TF complex at the site of injury [18,19].

Early during our in vitro studies of rFVIIa, I had observed that rFVIIa shortened the coagulation time in vitro also in the absence of TF but in the presence of phospholipids. This observation was based on results in the activated partial thromboplastin time (APTT, assay used for measuring the clotting time). In the APTT assay artificial phospholipid is used as a replacement for platelets. In this test, no TF is present, and thus, FVIIa should not be active in this test system. On the contrary, my observation indicated that FVIIa was also active in the presence of phospholipid/platelets without tissue factor, which was a surprise. The finding was presented and published at an international meeting in Brussels 1987, without arousing much interest [20]. These findings were confirmed shortly later by Telgt et al. [21]. Based on my original observations, I also suggested in my first publication on the use of rFVIIa in hemophilia patients in 1990 [3] that rFVIIa may not only bind to TF but also bind to phospholipids exposed on thrombin-activated platelet or on other cell surfaces. Finally, it was demonstrated that FVIIa bound to activated platelets independently of tissue factor [22].

It was clear to me that this should be investigated further. I realized that it was essential for us to know more about the mechanism of action of rFVIIa in hemophilia. Since our group at Novo Nordisk in Copenhagen was not allowed at that time, to use any resources on research in this area, I contacted Harold Roberts at Chapel Hill asking if his group would be interested in studying a potential interaction between rFVIIa and cell surfaces. During one of my visits to Chapel Hill and my discussion with Harold Roberts, it became clear that we needed a cell-based in vitro model for these studies. From the Chapel Hill group, Drs. Maureane Hoffman (experienced in working with cells) and Dougald M. Monroe became heavily involved with the work on establishing such a model. One of our biochemists, Marianne Kjalke, spent some months in Chapel Hill to participate in the work. From our laboratory in Copenhagen, we also contributed with reagents. This was the start of a close collaboration between our group and the hemostasis group in Chapel Hill, which led to several interesting results and a number of publications. An extensive discussion on the mechanism of action of rFVIIa will be given in Chapter 9, Mechanism of Action and Dosage.

The discoveries made represented a breakthrough in the area of hemostasis and have, indeed, met with much attention and interest. For me personally, the collaboration with the Chapel Hill group and with other research groups in the United States, e.g., Sam Rapaport's group in San Diego and later on Vijay Rao's group in Tyler, Texas, have been extremely rewarding. These collaborations also were helpful in the development of new hemostatic products in the research group at Novo Nordisk. Thus, we worked with mapping the so-called "tissue factor pathway inhibitor" (TFPI) first described by Sam Rapaport's group [23,24]. During the work with TFPI, we also studied its effect on hemostasis and were able to demonstrate, in animal models, that if TFPI was removed, bleeding decreased in animals with induced hemophilia. In a publication from 1995, we put forward the question as to whether this mechanism could be used in the treatment of hemophilia [25].

Another one of our projects at that time, for potential use as prophylactic thrombosis treatment in connection with surgery, was the development of a modified plasminogen molecule. It had long been known that plasminogen (its active form, plasmin, is the most effective enzyme for breaking down fibrin and an important part of the body's normal defense mechanism against the development of thrombosis) undergoes partial breakdown resulting in the exposure of another amino acid at the end of the molecule (lys-plasminogen), which binds more easily to fibrin and changes the fibrin structure to one which is more soluble [26]. My idea was to use this finding to develop a recombinant "lys-plasminogen" as part of our "antithrombosis" portfolio. Unfortunately, this project disappeared in the next big reorganization in 1995.

At the beginning of the 1990s, we also worked on producing recombinant aprotinin, an inhibitor to the fibrinolytic process. Novo Industri had earlier on sold aprotinin, manufactured from bovine material. Aprotinin or Trasylol, as Bayer's competitive product was called, became very popular during the 1980s—90s for decreasing hemorrhages during open heart surgery, which was becoming more common. However, Aprotinin had side effects affecting the kidneys. We wanted to modify the molecule to prevent its binding to cells in the urinary tract and thereby the development of acute kidney damage [27]. We had come a relatively long way in the development process, when the project was closed down (sold to Bayer) in connection with the large-scale reorganization of Novo Nordisk 1995.

5.5 THE DEVELOPMENT AND LICENSING OF rFVIIa 1990—96

The transfer of rFVIIa to the Biopharmaceutical Division also led to changed conditions for the continued clinical development, which was now transferred to the section for clinical development in the Division. Much competence in the hemostasis area was lost. Steven Glazer, who, in a very skillful way, had prepared the clinical plans for rFVIIa and developed a wide international network for the planned clinical studies, left the medical department and finally also, the company. He had not been able to complete work on the simplified protocol for the Compassionate Use program in the United States and had not managed to start the home treatment study that was planned since 1990.

Following the transfer of all further development of rFVIIa to the development and medical department of the division, my possibility to influence it was highly curtailed. On account of my close link to patient care and long-standing establishment in the care of hemophilia and hemostasis research internationally, I could, however, not be entirely eliminated from the continued development. For me the first half of the 1990s was an extremely frustrating period when I had to witness, more or less from the outside, how lack of resources and competence delayed developing rFVIIa to licensing status.

Regarding the continued work with the clinical development, it became more and more obvious that the allocated resources were insufficient. For example, the first study of hemophilia patients carried out in the United States started in 1989 was completed in 1992 but not reported until the end of 1993, because the Biopharmaceutical Division's Medical Department could not make resources available for the work. In May 1989, a calculation of resources necessary for the optimal continued development of rFVIIa had been sent to the management of Novo Nordisk. Here, it was emphasized that most of the allocated resources had been used to supply rFVIIa for hemophilia patients in acute need of treatment within the framework of the "Compassionate Use" program. They would not cover improvement of production technique and other preliminary work necessary to achieve licensing within a foreseeable future. However, it seemed that, amidst the general reorganization 1989—90, this appeal was not paid any attention.

However, the revised clinical plans for rFVIIa in 1991 still included a home treatment study with planned start in the last quarter of 1991. Also studies on patients with thrombocytopenia (low platelet counts) and reduced liver function were included. A pilot study of a few patients with thrombocytopenia had been carried out. Unfortunately, the beginning of the 1990s was dominated by endless discussions about the home treatment study. The protocol draft had been sent to the doctors who were to be included in the study and to selected employees at Novo Nordisk in December 1990. After a meeting with the potential participants in January 1991, the protocol was finalized and the study was planned to start in May1991. Unfortunately, there were a number of staff changes in the Medical Department of the Biopharmaceutical Division of Novo Nordisk during Spring 1991. For some obscure reason, the home treatment study was postponed even further. A new protocol was presented in October 1991, which required a renewed discussion with the participants. In addition, shortage of product delay in the completion of the form for patient reports and lack of resources in the affiliates were given as reasons for the various delays.

In January 1993, 3 years after the first protocol was presented, there was talk of an "almost completed protocol." In March 1993, at a meeting in Chapel Hill, the question of the home treatment protocol was once again taken up, obviously inspired by a new internal discussion at Novo Nordisk. A number of meetings took place at this time to establish the so-called "advisory boards" consisting of international experts. In my opinion, it was a mistake to appoint new members of these boards before each meeting. There was no continuity, and members had no opportunity to become acquainted with rFVIIa or with their colleagues at Novo Nordisk, which, I considered, lessened the influence of these advisory boards.

Furthermore, it was very time-consuming to contact different experts and arrange meetings suitable for everyone involved. However, still another "Advisory Board" meeting was held in Copenhagen on June 1993 followed by yet another one in July on the same year. The members of the "Advisory Board" strongly stressed the need for an immediate start of the study. For some reason, the report from both these meetings was delayed until December 1993. During the years 1993 and 1994 discussions about the home treatment protocol that was altered several times still were going on within Novo Nordisk, Copenhagen, and the American affiliate.

Twenty years later, it is interesting to read the intensive correspondence between the parties involved, especially those within Novo Nordisk. It clearly reflects the extreme frustration felt in face of the inability of the medical department to solve various problems. At one point, there was even discussion about simply discontinuing the further development of rFVIIa. In November 1993, the American affiliate noted that up to that point, they had seen 14 different versions of the protocol without a final decision being reached about the design of the study. Even now, it is hard to understand this total inability to go forward with the home treatment study during this period. However, finally, a home treatment study was started in the United States in spring 1995. It was reported as "ongoing" in the license documents sent in to EU in May 1994. The result of the study was published in 1998. It was obvious this period was the most frustrating period during the development of rFVIIa, also for me. In fact, this was the time I seriously considered to leave.

The tone between the medical department in Copenhagen and that of the American affiliate was at times resentful and bitter, in discussions not only about the home treatment study but also about the clinical development of rFVIIa in general. On one occasion, I offered to help with the data collection from the Compassionate Use study, which had been left out of action due to insufficient resources, and a lot of work was required to complete the material necessary for licensing. I was, however, told that it was not desirable. Instead, the obviously very frustrated American affiliate put forward the suggestion that I should abstain from holding lectures mentioning rFVIIa for 6 months, referring to the increased number of inquiries about and requests for rFVIIa in the acute treatment for hemophilia patients, which was noted after meetings and symposiums where I had been a guest lecturer. For me, these effects showed that the product was needed among patients and doctors, and should serve as a clear sign of a big market waiting for the product.

However, I realized that there was no hope that the project management would accept any suggestions from outside their group. Strict line control was employed, meaning an absolute control from the Biopharmaceutical Division's Medical Department. This did, however, not prevent me from having to take care of all the complaints from the members of the advisory boards concerning difficulties in agreeing on dosage, especially in the home treatment protocol. It also became my responsibility to sort out any strange laboratory results, which often turned up when going through patient data in the documentation work. I had nothing against this, but, nevertheless, would have appreciated if my suggestions had been taken into consideration from the beginning. Especially since they were based on actual clinical experience and the experience we had from developing of rFVIIa in the 1980s, before the merger and the large-scale reorganization took place. I thought that they would have been helpful.

During these tumultuous years, new clinical plans for rFVIIa aiming at the submission of documentation for licensing in Europe, the United States, and Japan at the end of 1994 were drawn up. The home treatment study planned to start in spring 1994, possibly only in Europe, as well as a study of rFVIIa for the treatment of acute hemorrhages in patients with a low platelet count (thrombocytopenia) was included. At this time, there was experience from a smaller uncontrolled study of patients with thrombocytopenia, carried out in the 1990s and published in 1996 [28]. In November 1994, the thrombocytopenia protocol was expected to be ready in the spring 1996. In the clinical development plans from 1993, it was also mentioned that development of rFVIIa for use in patients with an FVII deficiency and those with von Willebrand's disease was under consideration. At this point, rFVIIa had been used to treat both these conditions within the framework of the "Compassionate Use" program [29—31].

The application for licensing was submitted in March 1994 to the European authorities and included 112 patients in the Compassionate Use program ("emergency use") who had been treated with rFVIIa for life-threatening hemorrhages and acute surgery. Doses between 70 and 100 μg/kg body weight had been used. In total, the license documentation contained the results from four controlled studies (the dosage study carried out in 1989—90, the Compassionate Use study in 1988—95, the home treatment study, and the surgical study comparing two dosages. The two latter studies were described under the heading of "ongoing"). The documentation comprised 70 files. To prepare all these, a special building had to be rented for a limited period of time.

The rFVIIa was finally approved in the EU in February 1996 and was launched first in France (14th International Congress on Thrombosis of the Mediterranean League against Thromboembolic Diseases, October 14—19, 1996).

5.6 THE DEVELOPMENT OF PRODUCTION CAPACITY AND THE US LICENSE

As early as the 1980s an upscaling of production capacity for rFVIIa was commenced, and a factory with a production capacity of 5000 L was planned. Production was started in 1990, and a preliminary inspection by authorities was carried out in 1991. The pronouncement after the final inspection in August 1993 was that "Concerning FVIIa, Novo Nordisk is a superb and ethical entity" (Fig. 5.1).

In October 1991, Novo Nordisk met the FDA with the aim of discussing plans for a license application in the United States and the documents that would be necessary for an "expeditious review and approval of Factor FVIIa" for the restricted indication "treatment of hemophilia patients with inhibitors with life-threatening hemorrhages, when other treatment has proved insufficient." At this meeting, rFVIIa was presented in detail, both the characteristics of the molecule and details concerning production. A detailed plan for characterization of the product from the new production in 5000 L fermentation scale was also presented. The patients treated in the Compassionate Use program were presented in detail and effects, as well as risks, were discussed.

The meeting went very well indeed, and it can be seen from the report that the number of patients and bleeding episodes treated were considered sufficient. It was pointed out, however, that data from further

Figure 5.1 Inauguration of the rFVIIa factory in Kalundborg, Denmark, on April 1990. Author, Ulla Hedner, is in the center and to the right behind her is Kurt Pingel, the specialist in process chemistry.

patients treated during the application period should be submitted. Novo Nordisk had also submitted a suggestion for a reduced virus test program, which had been assessed by an external expert and which Dr. Maplethorpe at the FDA believed would be approved. After this meeting in October 1991, Dr. Maplethorpe complimented Novo Nordisk on a well-prepared and well-conducted presentation. We were also invited to a renewed meeting with the FDA as soon as possible to discuss plans regarding other indications for rFVIIa.

In a telephone call the next day, it was stated by representatives from FDA that they were going to carefully go through the IND (Investigation of a New Drug) sent in earlier to identify possible questions that could be answered by Novo Nordisk directly. Compliments from the other members of the FDA group who had been present at the meeting the day before were also forwarded. The representative of the licensing department of the American affiliate had informed Dr. Maplethorpe that we were already working on a document with information about the upscaled production, which he believed would be a help. From what I can see from my notes, the meeting we were invited to by the FDA took place on April 9, 1992. This meeting mainly focused on the documentation of the rFVIIa, which was being produced in the upscaled production. Even this meeting went very well and Novo Nordisk was complimented on the way the tests had been approached and carried out. The representatives from the FDA directly urged Novo Nordisk to start with the licensing process as soon as possible.

However, I can find no record of any follow-up meeting in the documents available to me. It was first found out in the summer 1993, more than 1 year later, that two people from Novo Nordisk had a private meeting with the FDA in October 1992, at which they said that they had only asked questions about the use of serum during the cell-fermentation phase in production. However, it became apparent that the meeting in October 1992 had been understood by the FDA as a meeting about rFVIIa in general and clinical questions in particular. The FDA had been informed about the meeting by a person at NIH (National Institute of Health) who had been contacted by Novo Nordisk. There had been no advance information about which questions the people from Novo Nordisk intended to raise.

Five representatives from the FDA took part in the meeting. It should be added that meetings with FDA usually were meticulously well prepared by Novo Nordisk. Accordingly, documents were sent to the agency long before the date of the planned meeting. In this document the questions Novo Nordisk wanted to discuss and a list of names of the people from the company who would participate were included. The meeting in October 1992 thus did not follow the normal routine used by Novo Nordisk. Therefore, it must have appeared strange to the FDA. According to a memo, written a few days after the October meeting by the Novo Nordisk participants, a question had been asked about what measures would be taken if a virus was found in a mammalian cell line. The FDA's answer was that it would be necessary to establish the origin of a possible virus, although it was proved to be eliminated later on in the production process. The FDA representatives also stressed that if a virus appears now and then, it is a sign that the production process is not under control.

During the winter 1993, the question of viruses was discussed backward and forward. The opinion at Novo Nordisk was that the problem most likely came from the bovine serum that was used initially in the cell fermentation. This was seen to be a serious problem, especially as other manufacturers of biological products, at this time, generally used serum-free processes in production. Novo Nordisk reckoned at the time on having a serum-free process within the next 5 years. Virus-validating processes were worked out, and shorter fermentation periods were introduced.

One of the difficulties in this phase was that the number of patients given treatment in the "Compassionate Use" program constantly was increasing, and it started to be a problem for the Novo Nordisk to produce the necessary large amounts at free of charge. Novo Nordisk in Denmark wanted first of all to send in a complete application for a license, with an indication limited to the treatment of life-threatening bleeding in patients with inhibitors against the coagulation FVIII/FIX, where existing treatment possibilities were ineffective. In this way it was assumed that the application could be processed more rapidly by the FDA. In Copenhagen, they wanted to build on the work and plans drawn up by the licensing department of Novo Nordisk in the United States at the beginning of the 1990s headed by Nathan Bloch. Those plans were discussed with the FDA in 1991 and 1992.

A disagreement between the Copenhagen group and the new licensing department in the US affiliate seemed to aggravate at this point, when the US group wanted to go for a type of "treatment IND." They argued that a complete production control would not be necessary if the "treatment IND" was chosen. Furthermore, the cost of the product could be debited, and they wanted to reduce the "Compassionate Use" program, a step that was firmly turned down by the Copenhagen group, including the medical department there. In Copenhagen they considered it to be detrimental to Novo Nordisk's reputation if these patients were suddenly refused treatment.

In August 1993, however, a protocol for "treatment IND" was sent from Novo Nordisk's American affiliate to the Medical Department of the Biopharmaceutical Division in Copenhagen, which indicates that a decision had been made to follow the line proposed in the United States. In this protocol, there was a new part about the criteria that should be used to allow patients access to rFVIIa for home treatment. In the middle of December on the same year, 1993, the licensing department in the United States telephoned Dr. Maplethorpe at the FDA to discuss "treatment IND" for rFVIIa. According to a memo, which I was permitted to copy, it stated that Novo Nordisk planned to submit a complete license application for rFVIIa "sometime during 1994," and that there was talk of "applying for a treatment IND at the same time." Dr. Maplethorpe stated that a "treatment IND" was not to be considered. Novo Nordisk ought to concentrate on finishing a complete license application. He also wondered whether Novo Nordisk had solved their virus problem and was informed that an external expert, an experienced inspector for the FDA, had been consulted and that details about production changes would be included in the license application.

At the beginning of December 1993, a decision was made by the Biopharmaceutical Division to put the "treatment IND" on hold. Later, during 1994, a resolution was sent from the "Medical & Scientific Advisory Council" in the United States (MASAC) stressing the great need for rFVIIa in the hemophilia area and urging the manufacturers of rFVIIa and the FDA to seriously work on finding ways to having the product approved and to making it accessible for clinical use.

In the beginning of 1994, it became obvious that there was a serious shortage of resources in several areas at Novo Nordisk. Serious complaints came from the preclinical area, where data were checked and errors in the documents corrected. The filling capacity of rFVIIa preparations had been affected by delays, because priority was given to growth hormone in the filling department. The project management announced that current rFVIIa studies risked being stopped because of the lack of product. Also, a license application could not be submitted without a reasonable stock of the finished product. At the same time, the importance of giving priority to licensing in Japan was underlined. The frustration over the large discrepancy between different plans and resources for rFVIIa in the various departments is abundantly clear in several documents from the beginning of 1994. At this time, the possibility of ever being able to get full licensing of rFVIIa was questioned from several quarters of Novo Nordisk.

Otherwise, spring 1994 was characterized by repeated discussions, not only about the lack of resources but also about reports of various potential problems. The drawbacks of taking such a long time for licensing of rFVIIa were clearly seen at this stage. During this period, people had been replaced and everyone had different ideas about whether to follow the guidelines from the beginning of the 1990s or those that had come into being later. As key persons in the decision-making process changed, new guidelines were suggested. For example, there was disagreement as to whether the report on potential contaminants should follow the recommendations from FDA 1990, which had been in use up to and including 1993.

In the summer 1994, a new strategy had been suggested by the US Novo Nordisk's licensing department. It was, thus, suggested that the results from the immunological investigations were to be kept separate from the final report and sent ahead of time to the FDA for assessment. Such sudden changes gave rise to time-consuming discussions as to what this procedure would imply for the European application and the time plan for the different Novo Nordisk departments. An even more complicated question was whether the biological method to determine if the product from the fully upscaled manufacturer was identical with that obtained from the 500 L scale could be changed. A more chemical method was advocated. However, although this method was more sophisticated, there was some uncertainty as to its validity.

On June 6, 1994, meeting was held at the request of the FDA. I was not present at the meeting, but I can see from the minutes that the FDA described it as a "re-acquaintance meeting" and not a "pre-PLA" meeting (meeting before the submission of an application for a license). The FDA representatives were especially skeptical about using a biological method to determine if the product of the fully scaled up production was identical with that of the 500 L production. It was pointed out several times that the sole use of a biological method was unsatisfactory. An assessment of clinical efficacy was called for. However, there were no questions about virus tests during production.

To summarize, the major questions at this FDA meeting were production—the biological test to verify the identity between the large-scale production and the former (500 L) production and how to determine if the clinical effect was identical. The FDA representatives seemed skeptical and underlined repeatedly that they were not certain that the production process was properly defined. They also announced that they intended to go through the clinical material in detail and decide whether Novo Nordisk needed to carry out further studies. This required internal discussions within the FDA. In a detailed answer, Novo Nordisk pointed out that the number of patients who responded positively to treatment with rFVIIa had been the same for several years, even when they had been treated with the large-scale product. This spoke against there being any great difference in the two production methods.

Additional work was done to validate a method for determining the carbohydrate content (glycopeptide index) in the rFVIIa product from the different production volumes to obtain a measure of the consistency between the two products. A final report on the glycopeptide index and its correlation to the biological method that had previously been used to ensure bioequivalence between the different production batches was presented at the beginning of August 1994.

During the following months, the more or less infected discussions between Novo Nordisk in Copenhagen and in the United States continued, regarding where the different tasks should be carried out and completed for the planned submission of the final license application (PLA) on September 30, 1994. This could seemingly not be done without a new FDA meeting focused on production facilities. Such a meeting was finally planned to take place on December 7, 1994.

After a long interval in the documents at my disposal, it was finally announced that an FDA meeting had been held on November 9, 1995, a year later than planned. It was a success. Novo Nordisk presented convincing proof that there was sufficient clinical data to send in the application. However, there was still work to do on the application for the approval of production facilities. It was recommended that this should be given the highest priority so that a complete application for a license could be submitted around New Year 1996.

The application for a license in the United States was finally sent in to the FDA on May 10, 1996, about 3 months after rFVIIa had been approved by the EU.

5.7 BLOOD PRODUCT ADVISORY COMMITTEE, 1996

Novo Nordisk was invited to a meeting with the Blood Product Advisory Committee (BPAC) in September 1996, specifically to discuss the clinical data on safety, dosage, and availability of the product, rFVIIa, during the ongoing examination of the submitted license documents. The FDA had expressed a special desire to discuss the patients treated in the "Compassionate Use" program. BPAC meetings had been launched by the FDA to spread information about their activities and to present certain products that were currently under consideration. A thorough presentation of rFVIIa had been sent ahead of time. As I was not a member of the project group or considered to have anything to do with rFVIIa at this time, I was only given Novo Nordisk's presentation for comment a couple of days before it was to be sent to the FDA, which gave me limited possibilities to comment.

At the BPAC meeting held on September 26, 1996, Novo Nordisk's coworkers were instructed by the director of Novo Nordisk's licensing department not to say anything during the meeting and they were not to answer any questions. It was apparent at the meeting that rFVIIa was strongly supported by the BPAC, while the FDA expressed doubts, as they were not convinced that Novo Nordisk had followed "rigid requirements." In my notes after the meeting, I deplored the fact that we had not been able to respond to the questions from Dr. Maplethorpe of the FDA directly during the meeting, as we had the answers and, in many cases, could have corrected or shed more light on the issues taken up by him.

The rFVIIa was finally approved under the name of NovoSeven in the United States in 1999. An application was sent to Japan in January 1995 and approved in 2000.

5.8 SOME FINAL REFLECTIONS

What was it that made me stay with Novo Nordisk in the 1990s with all the seemingly insurmountable problems that piled up against development in the area of hemostasis? In the middle of the decade, it was not even certain that rFVIIa would make it to the market despite its life-saving role for many hemophilia patients with inhibitors. The reorganization of Novo Nordisk in 1995 wiped out my entire field of research. All the hemostasis projects were abolished or sold off in connection with the reorganization. By pure chance, rFVIIa did not disappear from the company. The advice given by McKinsey, the consultant firm used in this reorganization, was to sell off all the hemostasis products and concentrate solely on growth hormone, apart from the diabetes area.

In this situation, there were mainly three reasons why I stayed and persisted in the rFVIIa project, although my influence was greatly restricted.

The most important factor was probably *the patient contact I had kept over the time*. The need for better treatment options for hemophilia patients with inhibitors was tremendous. I was contacted by colleagues around the world with questions concerning treatment in the most complicated patient situations. We worked together over the phone and with the help of rFVIIa managed to handle many extremely difficult cases. I learned a lot during these years, and as I have always been interested in the challenges presented by hemophilia treatment, this work helped me to endure all the frustrations at Novo Nordisk at this time. Furthermore, at this time I was convinced that rFVIIa was beneficial for hemophilia patients with inhibitors and I could not accept that this treatment possibility should be denied due to apparently insoluble developmental problems within the company.

Another important factor was *my, since the 1970s, deep-rooted interest in the role of FVII in the hemostatic process*. The availability of rFVIIa at this time enabled extended research on the mechanism of action of FVII/FVIIa and the FVII-dependent pathway of the hemostatic process. My work, together with a very competent research group, in the early development of rFVIIa was a source of great satisfaction. Our work led to several new basic understandings of the hemostatic process. During 1990s, when my research resources were drastically cut, I initiated close collaboration with the Chapel Hill research group, headed by Dr. Roberts. The work with this group eventually led to a modified model of the hemostatic process [19].

This scientific work played an important role in my decision to stay and continue working with rFVIIa and with other research activities directed at new products in the treatment of hemorrhagic and thromboembolic diseases.

A third area that also helped to make my work at Novo Nordisk interesting, despite all the difficulties, was *the annual symposia I initiated in 1986*. The first symposium was on thrombotic disease. The development of the low molecule heparin, Tinzaparin, made this my first project area at Novo Nordisk.

The idea of these symposia was to increase knowledge about hemostasis, including treatment and products in the field to my own research group and also to Novo Nordisk as a whole. As I had worked in the area of hemostasis for more than 10 years before I came to Novo Nordisk, I knew many international scientists and clinicians in the field. There was a great interest for these symposia, which were compared with the well-known Gordon Conferences, held once a year, and where preliminary research results were presented.

Apart from spreading knowledge about hemostasis at Novo Nordisk, I also found it important for our group to initiate contacts with different international scientists. It was quite impossible for us, at Novo Nordisk, with our limited resources, to cover the great need for research in this area. Both aims were achieved and the Novo Nordisk symposia were given the status of the foremost scientific meetings in the area of hemostasis.

The symposia were organized by myself and one of my coworkers in the marketing department responsible for the hemostasis area. I wrote personal letters of invitation to those we hoped to see as guest speakers, a job I used to do on the ferry between Dragör and Malmö on my way home around 6:00—8:00 p.m. We offered to pay for the flight to Copenhagen and for accommodation, but no fee was paid for the actual lecture, and during the first years, nobody asked for one.

The symposia activity developed to include one meeting on thrombotic diseases and one on bleeding disorders every other year alternately. As we collected more data about rFVIIa, the clinical part was expanded and, at the end of the 1990s, the symposia consisted of 1 day devoted to basic research and 1 day to clinical data. The symposia focusing on thrombotic diseases were discontinued when the heparin project was sold to Leo in the middle of the 1990s.

The structure with 1 day devoted to basic research and the second day to clinical data and the use of rFVIIa was kept until 2001. The marketing department then wanted more time for discussions on the use of rFVIIa, and it was decided that the so-called Copenhagen Meetings should be entirely devoted to this subject. It was further decided that future meetings should rotate between different affiliates instead of being concentrated to Copenhagen. To retain a forum for the discussion of research results in FVII—TF-dependent coagulation, the so-called Chapel Hill Meetings were initiated and were convened every other year as purely research meetings.

REFERENCES

[1] Macik BG, Hohneker J, Roberts HR, Griffin AM. Use of recombinant activated factor VII for treatment of a retropharyngeal hemorrhage in a hemophilic patient with a high titer inhibitor. Am J Hematol 1989;32:232—4.

[2] Roberts H.R. Letter to Ulla Hedner, May 15, 1988.

[3] Hedner U. FVIIa in the treatment of hemophilia. Blood Coagul Fibrinolysis 1990;1:307—17.

[4] Schmidt ML, Smith HE, Gamerman S, DiMichele D, Glazer S, Scott JP. Prolonged recombinant activated factor VII (rFVIIa) treatment for severe bleeding in a factor-IX deficient patient with an inhibitor. Br J Haematol 1991;78:460—3.

[5] Letter to Steven Glazer from the parents of the third hemophilia patient treated with rFVIIa in the Compassionate Use Program 1988.

[6] Stein S.F., Duncan A., Cutler D., Glazer S. Disseminated intravascular coagulation (DIC) in a haemophiliac treated with recombinant factor VIIa. In: 32nd Annual Meeting American Society of Hematology, Boston, MA; Nov 28—Dec 4, 1990; Abstract.

[7] Glazer S, Hedner U, Falch JF. Clinical update on the use of recombinant factor VII. In: Aledort LM, Hoyer LW, Lusher JM, Reisner HM, White GC, editors. Proceedings of the Second International Symposium on Inhibitors to Coagulation Factors. Advances in Experimental Medicine and Biology, Vol. 386. New York: Plenum Press; 1995. p. 163—74.

[8] Hedner U, Feldstedt M, Glazer S. Recombinant FVIIa in hemophilia treatment. In: Lusher JM, Kessler CM, editors. Hemophilia and von Willebrand's Disease in the 1990s. A New Decade of Hopes and Challenges. Amsterdam: Elsevier Science Publishers B.V; 1991. p. 283—92.

[9] Gringeri A, Santagostino E, Mannucci PM. Failure of recombinant activated factor VII during surgery in a hemophiliac with high-titer factor VIII antibody. Haemostasis 1991;21:1—4.

[10] McPherson J, Teague L, Lloyd J, Jupe D, Rowell J, Ockelford P, et al. Experience with recombinant factor VIIa in Australia and New Zealand. Haemostasis 1996;26 (suppl 1):109—17.

[11] Salaj P, Brabec P, Penka M, Pohlreichova V, Smejkal P, Cetkovsky P, et al. Effect of rFVIIa dose and time to treatment on patients with haemophilia and inhibitors: analysis of HemoRec registry data from the Czech Republic. Haemophilia 2009;15:752—9.

[12] Davie EW, Ratnoff OD. Waterfall sequence for intrinsic blood clotting. Science 1964;145:1310—12.

[13] MacFarlane RG. An enzyme cascade in the blood clotting mechanism, and its function as a biological amplifier. Nature 1964;202:498—9.

[14] Østerud B, Rapaport SI. Activation of factor IX by the reaction product of tissue factor and factor VII: additional pathway for initiating blood coagulation. Proc Natl Acad Sci 1977;74:5260—4.

[15] Østerud B, Rapaport SI. Activation of ^{125}I-factor IX and ^{125}I-factor X: effect of tissue factor and factor VII, factor Xa and thrombin. Scand J Haematol 1980;24:213—26.

[16] Rao LVM, Rapaport SI. Activation of factor VII bound to tissue factor: a key early step in the tissue factor pathway of blood coagulation. Proc Natl Acad Sci USA 1988;85(18):6687—91.

[17] Rapaport SI, Rao LVM. The tissue factor pathway: how it has become a "Prima Ballerina." Thromb Haemostas 1995;74:7—17.

[18] Roberts HR, Monroe DM, Hoffman M. Molecular biology and biochemistry of the coagulation factors and pathways of hemostasis. In: Beutler E, Lichtman MA, Coller BS, Kipps TJ, Seligsohn U, editors. Williams Hematology. 6th ed. New York: The McGraw-Hill Companies, Inc; 2001. p. 1409—34 [Chapter 112].
[19] Monroe DM, Hoffman M, Roberts HR. Platelets and thrombin generation. Arterioscler Thromb Vasc Biol 2002;22:1381—9.
[20] Hedner U., Lund-Hansen T., Winther D. Comparison of the effect of factor VII prepared from human plasma (pVIIa) and recombinant VII (rVIIa) in vitro and in rabbits. In: XIth International Congress on Thrombosis and Haemostasis, Brussels, Belgium, 1987, Thromb Haemostas 1987;58:270.
[21] Telgt DSC, Macik BG, McCord DM, Monroe DM, Roberts HR. Mechanism by which recombinant factor VIIa shortens the aPTT: activation of factor X in the absence of tissue factor. Thromb Res 1989;56:603—9.
[22] Monroe DM, Hoffman M, Oliver JA, Roberts HR. Platelet activity of high-dose factor VIIa is independent of tissue factor. Brit J Haematol 1997;99:542—7.
[23] Rapaport SI. Inhibition of factor VIIa/tissue factor-induced blood coagulation: with particular emphasis upon a factor Xa-dependent inhibitory mechanism. Blood 1989;73:359—65.
[24] Rapaport SI. The extrinsic pathway inhibitor: a regulator of tissue factor-dependent blood coagulation. Thromb Haemostas 1991;66:6—15.
[25] Erhardtsen E, Ezban M, Madsen MT, Diness V, Glazer S, Hedner U, et al. Blocking of tissue factor pathway inhibitor (TFPI) shortens the bleeding time in rabbits with antibody induced haemophilia A. Blood Coagul Fibrinolysis 1995;5:388—94.
[26] Wiman B, Wallén P. The specific interaction between plasminogen and fibrin. A physiological role of the lysine binding site in plasminogen. Thromb Res 1977;10:213—22.
[27] Lemmer Jr JH, Stanford W, Bonney SL, Breen JF, Chomka EV, Eldredge WJ, et al. Aprotinin for coronary bypass operations: efficacy, safety, and influence on early saphenous vein graft patency. A multicenter, randomized, double-blind, placebo-controlled study. J Thorac Cardiovasc Surg 1994;107:543—51.
[28] Kristensen J, Killander A, Hippe E, Helleberg C, Ellegård J, Holm M, et al. Clinical experience with recombinant factor VIIa in patients with thrombocytopenia. Haemostasis 1996;26:159—64.
[29] Ciavarella N, Schiavoni M, Valenzano E, Mangini F, Inchingolo F. Use of recombinant factor VIIa (NovoSeven[R]) in the treatment of two patients with type III von Willebrand's disease and an inhibitor against von Willebrand factor. Haemostasis 1996;26:150—4.
[30] Ingerslev J, Knudsen L, Hvid I, Tange MR, Fredberg U, Sneppen O. Use of recombinant factor VII in surgery in factor VII deficient patients. Haemophilia 1997;3:215—18.
[31] Hedner U, Ingerslev J. Clinical use of recombinant FVIIa (rFVIIa). Transfus Sci 1998;19:163—76.

CHAPTER SIX

Legal Issues Regarding the Use of rFVIIa in Hemophilia Patients With Inhibitors

Contents

6.1 "ORPHAN DRUG ACT"

During the first years of the 1980s, the need to develop products for the treatment of small patient groups with rare disorders had been brought to attention in the United States. To compensate pharmaceutical companies somewhat for the risk of never recovering the costs invested in drugs for these patient groups, a law was passed ensuring a prolonged patent-protection period and exclusive marketing rights, the so-called "Orphan Drug Act" [1]. A drug that had been approved according to this "Orphan Drug Act" was protected against applications for new patents within the same area of disease for 7 years. The possibility of receiving economic help for the costs of clinical testing carried out in the United States was included. This law was passed January 4, 1983.

I was made aware of the law during one of my flights between Copenhagen and the United States in the middle of the 1980s. The passenger in the seat beside me was a lawyer from New York City, who had been in Gothenburg, Sweden, on business. When he got to know that I worked with the treatment of hemophilia, he asked me if I knew about this relatively new law, which, of course, I did not. He offered to send me the relevant documentation and did so in October 1985. These documents then formed the basis of our work to get rFVIIa accepted as an "orphan drug." Based on a document describing hemophilia, its

Treating Life-Threatening Bleedings.
DOI: http://dx.doi.org/10.1016/B978-0-12-812439-0.00006-X

treatment, and the reason why improved treatment was required written by me, Novo Industri A/S requested designation of Factor VIIa, recombinant DNA origin in 1987, and Orphan Drug status was granted to rFVIIa in January 1988 [2].

6.2 PATENT RIGHTS (INTELLECTUAL PROPERTY)

In September 1983, I received a letter from Walter Kisiel telling me that it had just come to his attention that Baxter-Travenol, one of the largest American pharmaceutical companies in the hemostasis area, had applied for a patent regarding a method and product for the treatment of patients with inhibitors against coagulation factors. The application was sent in and registered on May 11, 1981, and covered a product containing factor VII and factor VIIa, as well as other coagulation factors such as factors IX, IXa, X, and Xa. Walter Kisiel advised me to contact a lawyer at Novo to discuss the possibility of contesting the patent. He immediately realized that there was a large risk that Baxter-Travenol, with the help of this patent, could prevent the development of recombinant FVIIa for many years to come, if no one contested this patent application. For Walter Kisiel and myself, it seemed unbelievable that Baxter could obtain patent on something we had presented at a scientific meeting more than a month before the application and before I had any contact whatsoever with Novo Nordisk. At that time, I was employed by Lund's University in Sweden.

I had, namely, at a meeting in Rome on March 30–31, 1981, more than a month before Baxter had submitted their patent application, presented our work on producing completely pure FVIIa from human plasma [3]. In the same presentation, I had also shown that the injection of our preparation in healthy dogs did not give rise to any changes in the plasma coagulation factors, changes which indicated a general activation of the hemostatic system. A general activation of the coagulation mechanism may give rise to the clotting in the circulating blood and thus thromboembolic side effects occur. Such side effects are seen after the injection of activated prothrombincomplex concentrate (aPCC), which I had demonstrated earlier [4,5].

In my presentation in Rome, I stated quite clearly that I intended to treat a hemophilia patient with inhibitor against FVIII in conjunction with a bleeding episode in the near future. This treatment took place on

April 24, 1981 (see Section 2.4). Baxter's patent covered aPCC products, and one section in the application especially mentioned the content of a defined amount of FVII and FVIIa in these products [6]. This would quite clearly prevent the use of rFVIIa in the treatment of hemophilia patients with inhibitors, placing a serious obstacle in the way of Novo's development of an rFVIIa preparation.

It turned out that a European patent application had also been submitted by Baxter Travenol on May 26, 1982, and published in January 1983 [7]. Therefore, it was important that Novo investigated the matter further. This was the start of long drawn out and demanding work with the patent, which did not end until 1992, when we lost the case against Baxter in the European Patents Court, despite having been able to prove certain irregularities in the handling of the patent.

We had to prove that my contribution in Rome on March 30, 1981, had made public the possibility of using FVIIa for treating hemophiliacs with inhibitors, before Baxter submitted their patent. The first assessment of Novo's patent department on June 6, 1984, based on the applications Baxter had sent with priority and dated June 25, 1981, was that Baxter would probably not be granted patent because of its lack of invention value and, possibly, also because of lack of newsworthiness.

It was decided that Novo should monitor the European application. As I had made my contribution during the Rome meeting during a discussion, it had not been published in the book with the preregistered lectures. I had kept my manuscript from the presentation but would very much like to have it confirmed by someone in the audience. There was obviously no recording of the session, but two of the participants ensured that they had heard me speak about FVIIa as the active principle in "bypass" therapy of hemophilia patients with inhibitors and that I had also spoken of dogs, as well as the use of FVIIa in humans [8].

During the years that followed, we wrote numerous submissions to both the American and European patent authorities, repeatedly showing that (1) the production of the preparations containing FVII and FVIIa, separately or together in combination with other factors in the prothrombin complex, had been described in several articles before Baxter had applied for patent and that it could not therefore be considered a new invention and (2) such a preparation lacked news value, as I had already in 1981 presented data confirming the use of FVIIa in the treatment of hemophilia. Negotiations between Novo and Baxter went on for several years during which, as far as I can see from documents,

Novo Nordisk emphasized that they would preferably like to come to an acceptable agreement instead of taking legal proceedings regarding the validity of the patent.

In Europe, Novo Nordisk thought they had a good chance of winning a patent case based on my presentation in Rome in March 1981. However, at the trial in the European Patents Court, it proved impossible to convince the jury of my credibility in this connection. Instead they listened to Baxter's barrister, who depicted me as an unknown woman with a dubious research background, who had talked about dogs. He meant that the auditorium at the Rome meeting could not be expected to attach any importance to an interjection from me, and that my presentation, therefore, could not be expected to have led to public knowledge about our results with pure FVIIa and our intentions to use it in the treatment of hemophilia patients. Our only external witness, one of the scientists who had been present at the Rome Meeting, also faltered when faced by the elaborately skillful presentation and interpretation of my contribution in Rome by Baxter's barrister. The trial in Munich was a thoroughly unpleasant experience for me, confirming my earlier suspicions that the judicial system is not to be trusted. Present in the court were also representatives from Baxter who knew very well that there were a number of obvious irregularities behind the patent in question. Novo Nordisk appealed the outcome of the trial, but after an agreement with Baxter had been drawn up at the end of the 1990s, the appeal was withdrawn.

Thus, Baxter were given both the European and the American patent for products containing FVII, activated FVII (FVIIa) and other coagulation factors in the prothrombin complex for the treatment of patients with inhibitors against coagulation factors. An agreement was eventually reached between Novo Nordisk and Baxter in which Novo Nordisk was to pay a certain license fee to Baxter.

REFERENCES

[1] Public Law 97-414 [H.R. 5238]; January 4, 1983. House Report No. 97-840. Congressional Record, Vol. 128 (1982).
[2] Request for designation of Factor VIIa, recombinant DNA origin as an orphan drug. December 15, 1987.
[3] Mariani G., Russo M.A., Mandelli F., eds. Activated prothrombin complex concentrates. Managing hemophilia with factor VIII inhibitor. In: Proceedings of the International Meeting on Activated Prothrombin Complex Concentrates: The State of the Art of Managing Hemophilia With Factor VIII Inhibitor. Rome, March 30–31, 1981: Praeger Publishers, 1982.

[4] Hedner U, Nilsson IM, Bergentz SE. Various Prothrombin complex concentrates and their effect on coagulation and fibrinolysis in vivo. Thromb Haemost 1976;35:386—95.
[5] Hedner U, Nilsson IM, Bergentz SE. Studies on thrombogenic activities in two prothrombin complex concentrates. Thromb Haemost 1979;42:1022—32.
[6] William R. Thomas, United States Patent Application Number: 277,469. Therapeutic method for treating blood-clotting defects with factor VIIa. Filed on June 25, 1981, Baxter Travenol Laboratories, Inc., Deerfield, IL.
[7] William R. Thomas, European Patent Specification Application Number: 82902136.9. Composition based on the hemostatic agent factor VIIa and method of preparing same. Filed on May 26, 1982; Publication Number: 0082 182; Proprietor: Baxter Travenol Laboratories, Inc., Deerfield, IL.
[8] Over and Bonno Bouma, Affidavits from two of the participants in the Rome meeting, March 30—31, 1981, September 11, 1990.

CHAPTER SEVEN

Treatment With rFVIIa in Malmö (1996—99)

Contents

7.1 PART-TIME EMPLOYMENT AT THE COAGULATION CLINIC IN MALMÖ (1996—90)

During the clinical development of rFVIIa, it had become clear to me that we needed to broaden our knowledge about dosage and pharmacokinetics. As I mentioned in Chapter 5, The further Use and Development of rFVIIa (1989—1996), it was clear that the dosages given in the licensing documents need to be modified. At this time, the interest for rFVIIa from the Marketing Department at Novo Nordisk was still limited. There were strong doubts about the saleability of rFVIIa and therefore the interest for new, broadened studies of its clinical effect was lukewarm. In this connection, although without direct relation, I was asked by the Coagulation Clinic in Malmö if I would consider working there, for example, 2 days a week. The idea was that I would be supporting the research work at the clinic.

I saw at once that a job at the clinic in Malmö provided me with the possibility of using rFVIIa in the treatment for hemophilia patients with inhibitors. It would provide me with the opportunity of working with modified dosage. The results from the study of rFVIIa in home treatment which was published in 1998 showed that the dose used, 90 µg/kg, was not always sufficient to stop mild to moderate bleeding with a single injection, which is necessary for an optimal effect [1]. Provided that treatment is started early on in the process, more than one injection should

Treating Life-Threatening Bleedings.
DOI: http://dx.doi.org/10.1016/B978-0-12-812439-0.00007-1

not be necessary. As a result of the total reorganization of Novo Nordisk in 1995, when functional divisions were introduced instead of having different functions, for example, pharmacology, analysis, and production development combined in the area of therapy, my area of hemostasis had disappeared. It was, therefore relatively easy for me to reduce my working hours at Novo Nordisk and, instead, spend 2 days a week at the hemophilia clinic in Malmö. At Novo Nordisk, my task as "scientific director" for rFVIIa was to act as an adviser and, to some extent, as an initiator of development in dosage and widened indication areas. I shall return to this in Chapter 8, The Launching and Uses of rFVIIa.

At the clinic in Malmö, I mainly dealt with routine work, investigating patients with bleeding and thrombotic diseases. It was a good way for me to update my knowledge in areas other than hemophilia. Primarily, I did not deal with the usual hemophilia treatment but became quickly involved in complicated patients with inhibitors. It gave me the opportunity of using rFVIIa in practice which was very valuable. It was in this way that I came into contact with my third "key patient" in the development of the treatment of hemophilia. The first one was a boy, treated in 1981 with pure FVIIa, produced from human plasma, while I was still at the hemophilia center in Malmö, and the second was the first patient treated with rFVIIa in conjunction with a knee operation in 1988.

7.2 THE THIRD "KEY PATIENT"

The third "key patient" was originally from Australia and had been sent to the clinic in Malmö to undergo immune tolerance treatment according to the so-called Malmö model [2]. From the very beginning, he had a complicated history of illness with three cerebral hemorrhages during his first 3 years of life. He then developed inhibitors against FIX, which complicated treatment. During 1991–93, he underwent immune tolerance treatment in Malmö using high FIX concentrate doses.

The first immune tolerance treatment was started at the end of October 1991 and was, from the beginning, complicated by hemorrhages in both forearms in connection with the insertion of a catheter, necessary to ensure durable access to the vascular system. From October 1991 to April 1993, the patient underwent three immune tolerance treatments without any long-lasting effect on inhibitor levels. In May 1993, I had my first contact with the patient's mother who contacted me at Novo Nordisk on her own initiative. As I was away at that time, she was referred to the medical section where they concluded that she had lost faith that immune tolerance treatment could help her son, although it had succeeded in helping his twin brother.

It appeared that the patient's mother had heard from Australia, her home country, that rFVIIa was used for the treatment of hemophilia B patients with inhibitors. She had heard my name mentioned in this connection and wanted to talk to me. Urged by the medical section at Novo Nordisk having received her telephone call, I phoned her at the beginning of May 1993. She asked if her son could be given access to rFVIIa for the treatment of acute bleedings and I explained that Novo Nordisk were not able to provide any treatment. However, I informed her that Novo Nordisk had an international program, the "Compassionate Use Program" in which hemophilia patients with inhibitors could have access to rFVIIa for the treatment of "life- and limb-threatening bleedings."

Access to this program was mediated by Novo Nordisk's affiliate in Australia in collaboration with the Australian hemophilia doctor responsible for the patient. I advised her to contact him. Later on in July 1993 I was given the opportunity to comment on an application from Novo Nordisk in Australia and Dr. Lammi at the Hematology Unit, Children's Hospital, Sydney, to include one of the twin boys in Australia in the "Compassionate Use Program." The application was approved in July 1993. In August 1993, the family decided to move back to Sweden and requested help to transfer the approval to Sweden. The Medical Department at Novo Nordisk in Copenhagen advised them to contact the Hemophilia Centre in Malmö if the need arose.

In March 1994, the boy was treated for the first time with rFVIIa. In the meantime, he had had a cerebral hemorrhage (August 1993) and several gum and nose bleedings, which had been treated with Cyklokapron (Tranexamic acid, which inhibits the dissolution of formed thrombi). After a month-long bleeding from the nose and mouth had affected him, in March 1994 he was finally treated with rFVIIa through the "Compassionate Use Program" for which he had been accepted in July 1993. One wonders why, in connection with the severe bleedings during the first immune tolerance treatment in October 1991, there had been no application for treatment with rFVIIa according to the "Compassionate Use Program" that started in 1990—91, and why he was not been treated with rFVIIa in connection with the cerebral hemorrhage in August 1993, when he had access to this program via Novo Nordisk in Australia.

A year after the first treatment with rFVIIa, the patient was given his second treatment in connection with a severe knee joint bleeding. According to the existing documentation, the dosage at this time was 120 μg/kg body weight and was administered every other hour for approximately 24 hours, after which the dose was increased to 180 μg/kg every other hour for a further 12 hours. The effect was judged to be poor and the doctor in charge drew the conclusion that treatment of the knee joint bleeding with rFVIIa had no more effect than if no treatment had been given at all. It should be emphasized that the effect of hemophilia treatment is considerably better if started directly at the first signs of a hemorrhage. In the actual case the knee joint hemorrhage had gone on for some time (the patient was in severe pain) before treatment with rFVIIa was started.

During the latter part of 1995, a new immune tolerance treatment was discussed and planned to take place in January 1996. For some reason this did not happen. In October 1996 the patient was given his third treatment with rFVIIa in connection with an acute hemorrhage in the same knee that had been treated previously. Under the cover of rFVIIa (160 μg/kg, the first three doses every other hour and a longer interval thereafter) the knee was punctured and a large amount of blood suctioned out, after which radioactive yttrium was injected into the joint. This type of treatment for repeated bleedings in the same joint is an accepted treatment to decrease inflammatory changes in joint tissue [3,4]. However, for the optimal treatment of a joint hemorrhage, treatment should be started at an early stage before large amounts of blood have accumulated in the joint.

At this time, in autumn 1996, the patient was regarded as more or less resistant to treatment with rFVIIa. He had developed contractures in both knee joints and was wheelchair bound. Besides this, he had repeated bleedings in other joints as well as from the nose and soft tissues. He had access to rFVIIa as home treatment but had got the impression that it did not work and was, therefore, not especially interested in starting treatment immediately after he noticed symptoms of an incipient bleeding. The dose prescribed was 160 μg/kg. The possibility of trying to administer rFVIIa as a continuous infusion in connection with hospital care was also discussed.

When I started my part-time work at the coagulation clinic in Malmö, autumn 1996, besides the routine work, I took the initiative to start a pharmacokinetic study of the six hemophilia patients with inhibitors in Sweden who would be eligible for treatment with rFVIIa. In collaboration with the other hemophilia specialists in Sweden a study was carried out measuring different pharmacokinetic parameters by taking blood tests up to 6 hours after an injection. The ideal pharmacokinetic study should include blood tests taken over the first 24 hours, which was impossible under the prevailing circumstances. There was no way of making the patients stay overnight or longer. The analyses were carried out in the Analysis Section at Novo Nordisk by one of the technicians from the previous hemostasis section and showed quite clearly that rFVIIa had a higher clearance rate up to three times in children than that found in adults [5]. It was seen that the patient who had been classified as unresponsive to rFVIIa treatment had the fastest clearance rate. This supported my idea that the Australian boy needed a much higher dose than the one recommended in the licensing documents for rFVIIa. The higher clearance rate in children was later confirmed [6,7].

It was not until the middle of August 1998 that I met the boy at the hospital in Malmö although I had seen him in 1997 on the street outside the house where I lived at the time, as his family were our neighbors. I had felt upset and distraught at the thought that a hemophilia patient with inhibitors was sitting in a wheelchair in Sweden in 1997. I had invested so much energy and drive into producing a drug that would improve the situation for these patients.

In the middle of August 1998, I came across him a Monday morning at the Children's Hospital in Malmö. He had been hospitalized since Saturday after developing an elbow bleeding on Friday which was treated with two bolus doses of 160 μg/kg of rFVIIa with a 6-hour interval. On arrival at the hospital, he was in severe pain and was given a continuous infusion with rFVIIa 20 μg/kg per hour. As the pain did not abate, the patient was given morphine. The swelling increased over and under the elbow joint. When his mother came to visit by Sunday morning, the swelling had increased even more and the boy was in very severe pain. She had brought with her the rFVIIa which she had had at home and, on her own initiative, gave him two bolus doses (160 μg/kg) with a 2-hour interval. During the day, the swelling seemed to have stabilized and had not increased the next day. The pain also abated the same day.

On the basis of a meticulous documentation of the course of this elbow hemorrhage, and the previous pharmacokinetic results that showed that this special patient had three times faster clearance rate of the injected rFVIIa than in a normal adult, I raised the possibility with the head of the coagulation unit of increasing the boy's dose of rFVIIa. I suggested that the dose be doubled to make sure a hemorrhage should come to a complete stop with a single injection. Besides that, I asked permission to speak to the boy's mother to explain how important it was to start treatment at the first sign of any symptom.

A new plan for the treatment of this patient with rFVIIa was drawn up, with a bolus dose of 320 μg/kg to be given immediately when the first symptom appeared at home. In addition to this, the importance was stressed of measuring the actual size of the joint swelling to facilitate the assessment of the treatment effect. The mother was also requested to keep a careful account of every hemorrhage.

The next hemorrhage, which was a soft-tissue bleeding in the left forearm, happened a fortnight later. This bleeding was treated at home with the prescribed higher dosage (320 μg/kg: 2 × 4.8 mg). The boy was hospitalized overnight for tests and observation. After a night's uninterrupted sleep, he was sent home again. During the remainder of autumn 1998, the patient was treated with the same dose (320 μg/kg) for every bleeding (elbow, nosebleed, soft-tissue bleeding below the shoulder, ankle, knee, and tooth loss). The treatment was given at home and only on one occasion (tooth loss) was it necessary to administer two doses of 320 μg/kg to completely stop the bleeding. During this period the use of rFVIIa decreased successively from two to one dose a week. From February 1999 the total amount of rFVIIa decreased further to one to two doses at 320 μg/kg a month. Thus, by treating each bleeding episode immediately and with an adequate dose, the number of bleedings was markedly reduced. During this time, the boy could participate in regular physiotherapy that helped strengthen his muscles. However, the patient still had double-sided knee contractures, which made him dependent on a wheelchair and crutches.

At this point, I began to think about the possibility of using what Dr. Åke Ahlberg called "late measures" for the treatment of pronounced handicaps. Dr. Åke Ahlberg was the orthopedic surgeon who had treated several hemophilia patients with such problems in the 1960s, when Sweden began to have access to treatment with FVIII/FIX concentrate, making it possible to perform orthopedic surgery. These "late measures" comprised continuous traction and operation. Continuous traction was especially used to treat contractures that had been established for a long time, for example, in knee joints, and which had not responded successfully to normal physiotherapy (Fig. 7.1).

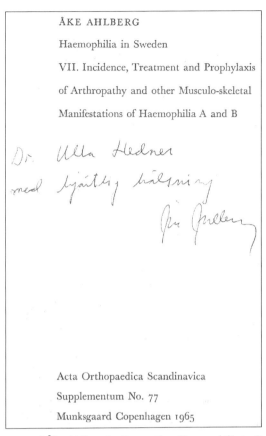

Figure 7.1 The cover of Åke Ahlberg's dissertation *Haemophilia in Sweden*, defended at the University of Lund, Sweden, in 1965.

The technique implied that the patient's entire leg was put in a plaster cast, which was cut up below the knee. With the help of a hinge on the front of the knee and a system of pulleys, the knee joint was stretched by slowly twisting a rope (once a day), thus diminishing the stretch defect. The process must be painless for the patient and therefore takes place over a long time. With the help of this technique stretch defects of as much as 90° have been eliminated and the result has lasted for several years. The treatment must, however, be carried out under the regular administration of a coagulation factor preparation to avoid joint bleeding during the stretching process [8].

The hemophilia patient in question had stretch defects of 30° and 45° in his knees and we had learned that a dose of 320 µg/kg rVIIa effectively stopped both his joint and soft-tissue bleedings. I thought that it should be possible to eliminate his stretch defects with the help of these techniques, which had successfully been used on many hemophilia patients with stretch defects up to 75°−90°. Nine such patients, who had all started by being wheelchair bound or only able to walk with the help of crutches, had been presented in Åke Ahlberg's dissertation in 1965 [8]. During traction treatment they had been given moderate doses of coagulation factor preparation. Despite the fact that the level of FVIII in plasma was often not raised more than a percentage, no bleeding complications had been noted and all the patients have managed exceptionally well later when they were free of their knee contractures.

Based on these results, which I could access from the original patients' records, I suggested that we should let our current patient with hemophilia B with inhibitors undergo traction treatment. This was met with very mixed feelings by the staff at the Hemophilia Clinic of Malmö, involved in the care of the patient. The orthopedic surgeon in charge was especially negative and was supported by the hematology physician. They were of the opinion that traction treatment could damage the knee joints. No attention was paid to the experience of such treatment published 30 years earlier by Åke Ahlberg where all hemophilia patients he treated were living examples of a successful result.

Besides this, the orthopedist expressed a categorical opinion that an operation was out of the question for the patient unless the coagulation expertise could guarantee absolute hemostasis. This was obviously interpreted by the coagulation experts as a demand that the patient's inhibitors must be totally eliminated, which, of course, seemed to be impossible to be achieved after three unsuccessful and complicated immune intolerance treatments. It was obviously difficult for the experts involved to take in the information that rFVIIa had proved to be an effective hemostaticum. It made to perform the major orthopedic surgery possible without bleeding in patients with severe hemophilia complicated by inhibitors [9,10].

Likewise, it was difficult to acknowledge the successful results achieved by traction treatment during the 1960s−70s in patients who, when followed up to the 1990s, were seen to have had excellent long-term help of the treatment. Thus, the situation in November 1998 seemed to have come to a standstill. It was abundantly clear that the desired effect could not be reached with the sole help of physiotherapy. In January 1999, the coagulation experts (including a pediatrician) decided that traction treatment for the patient's knee joints should be avoided.

Parallel to this, I suggested to the orthopedic surgeon that we should go through the nine patients in Åke Ahlberg's dissertation from 1965 and call them in for a clinical examination to document their progress on a long-term basis after traction treatment in the 1960s–70s. I went through their clinical data in their files myself, but unfortunately won no support for the follow-up study, which would have given us facts about the long-term lasting effect of traction treatment on severe knee damage in hemophilia patients. However, I had difficulty to let go of my thoughts about our hemophilia patient with inhibitors who had responded excellently to the treatment with rFVIIa and continued to sit in a wheelchair in 1999 with stretch defects in his knee joints. In connection with the treatment of this patient, I had lost my part-time employment at the hemophilia clinic in Malmö and therefore, in principle, was excluded from further discussion about his future treatment. It seemed that he was only to be given physiotherapy, which led to a moderate improvement.

The patient's mother, however, finally demanded a "second opinion" of the boy's treatment and as we were still neighbors in Malmö she asked me privately for a suggestion. I recommended her to contact Professor Harold Roberts, head of the Center for Thrombosis and Hemostasis in Chapel Hill at the University of North Carolina. Dr. Roberts had experience of rFVIIa from the beginning, that is, from the end of the 1980s. Among other things, he had treated the second patient globally with rFVIIa in 1988. I sent a summary of the patient's case history to Chapel Hill and they promised to see the patient and his mother for a consultation in June 1999.

The orthopedic surgeon at Chapel Hill thought that the boy should immediately be given traction treatment. As the people in Malmö were still dubious and wanted to further postpone the treatment, I took the initiative of suggesting to Novo Nordisk in the United States that they contribute by providing the rFVIIa to cover the treatment. It was clear that the treatment recommended at Chapel Hill would take a long time with rFVIIa to avoid hemorrhages during the intensive physiotherapy that was considered necessary. I emphasized to the Novo Nordisk Affiliate in the United States that this would be the first time rFVIIa would be used as prophylaxis.

At this time, it was not clear that rFVIIa would function as prophylaxis, that is, be able to prevent the occurrence of bleeding and not just stop ongoing bleeding. Prophylaxis with FVIII/FIX preparations had been used in hemophilia A/B patients since the 1950s based on the theory that the addition of FVIII/FIX to a plasma level of 1%–5% would transform severe hemophilia into a moderate form [11,12]. As rFVIIa disappears from the bloodstream much faster than FVIII/FIX, it was not certain that it would work as a prophylactic treatment.

However, we had seen that treating our current hemophilia B patient with inhibitors with a higher dose of rFVIIa had an effect even if the plasma level of FVII had returned to its initial value. We already knew that rFVIIa, directly after injection, has a larger distribution volume than that corresponding to the immediate plasma volume [13]. This is a sign that rFVIIa passes through the vessel wall and is also distributed in the extravascular space. These observations gave rise to new thoughts that perhaps rFVIIa not only was effective in forming a complex with tissue factors around the local damage to the vessel wall but may also form an extravascular complex, creating a prolonged effect by immediately stopping small microbleedings [14]. I believed it would be extremely valuable if we could show that our patient could undergo a longer traction treatment of his severe knee joint contractures with the protection of rFVIIa. This would be an incentive to further develop rFVIIa for prophylactic use and thus be able to provide prophylaxis even for hemophilia patients with inhibitors.

To everyone's delight, the Novo Nordisk USA Affiliate could see the advantages of such an undertaking and promised to provide rFVIIa for the treatment in Chapel Hill. I continued to relate to the Chapel Hill physicians our experiences with severely handicapped Swedish hemophilia patients, who had been treated in Malmö and were managing exceptionally well in their daily occupations. I actually knew all nine of them with the most pronounced defects, and, thus, had personal experience of how well it had turned out for them.

The traction treatment at Chapel Hill finally started in the middle of May 1999 with a dose of rFVIIa (the higher dose which had proved to be hemostatically effective in this patient) several times daily for the first days, when stretching with the help of weights was used, considered to have the greatest risk of causing bleeding. After this, the treatment was continued with an increasing interval between doses until it finally was given once a day during the rest of the rehabilitation period. The group at Chapel Hill judged the treatment with rFVIIa to be successful and after 2 months they planned to send the boy back to Sweden.

The pediatrician in charge meant that in the future, it should be possible to keep the boy hemorrhage free with the help of rFVIIa. During the entire traction treatment he had only two moderate bleedings that demanded extra rFVIIa. Both of them had been induced by trauma (Fig. 7.2).

This successful traction treatment (Fig. 7.3) was published in 2001 [15].

Figure 7.2 The patient who went through a traction treatment of both his knee joints under the cover of rFVIIa in Chapel Hill, NC, in September 1999. The photograph was taken during the traction treatment when his legs were in plaster cast.

7.3 WHAT DID THIS PATIENT TEACH US?

The treatment of the severely ill hemophilia patient from Australia taught us the following:

1. The unsuccessful response of a patient to a dose of rFVIIa could be changed into an excellent response if the dose was increased.
2. The dose necessary for an adequate response in an acute bleeding episode seemed to be dependent on the patient's clearance rate for rFVIIa. The initial dose of rFVIIa must be large enough to secure the formation of a stable fibrin plug with a dense network at the site of damage to the

Figure 7.3 The same patient as in Fig. 7.2 after he was back to Australia in 2000 and able to walk without crutches or other support. Both photographs were taken by his mother.

vessel wall. A fibrin plug with a dense network formed in the presence of a rapid, high thrombin concentration is more resistant to the premature dissolution of a fibrin plug. A sustainable fibrin plug is necessary for permanent hemostasis. It is known that children often have a higher clearance rate than adults, which we demonstrated in our third "key patient" and also in other children [5–7]. By increasing the dose of rFVIIa, exceptionally good results were obtained by one injection in connection with mild to moderate bleeding, provided the treatment was started immediately after the onset of the symptoms.

3. By using the higher dose of rFVIIa, identified in the treatment of moderate joint and muscle bleeding, the pronounced knee joint contractures could be treated successfully with orthopedic traction treatment without complicating bleedings. During the main part of the traction treatment, it was found to be sufficient with one dose of rFVIIa per day. This indicated that rFVIIa could be effective in

prophylaxis for patients with inhibitors. This had been my aim since I started working on the development of rFVIIa for the treatment of patients with inhibitors, that is to say, to make the treatment for these patients equivalent to that of hemophilia patients without inhibitors.

The encouraging results from "the third key patient" caused me to start thinking and searching for an explanation of the prolonged effect of rFVIIa in preventing bleedings to occur despite a rapid clearance rate from the blood. It also gave the basis for further studies on the dosage and the prophylactic use of rFVIIa, which I shall return to in Chapter 8, The Launching and Uses of rFVIIa.

REFERENCES

[1] Key NS, Aledort LM, Beardsley D, Cooper HA, Davignon G, Ewenstein BM, et al. Home treatment of mild to moderate bleeding episode using recombinant factor VIIa (NovoSeven) in haemophiliacs with inhibitors. Thromb Haemost 1998; 80:912—18.

[2] Nilsson IM, Jonsson S, Sundqvist S-B, Ahlberg Å, Bergentz SE. A procedure for removing high titer antibodies by extracorporeal protein-A-Sepharose adsorption in hemophilia: substitution therapy and surgery in a patient with hemophilia B and antibodies. Blood 1981;58:38—44.

[3] Ahlberg Å, Pettersson H. Synoviorthesis with radioactive gold in hemophiliacs. Acta Orthop Scand 1979;50:513—17.

[4] Rodriguez-Merchan EC, Caviglia HA, Magallon M, Perez-Bianco R. Chemical Synovectomy vs. radioactive synovectomy for the treatment of chronic haemophilic synovitis: a prospective short-term study. Haemophilia 1997;3:118—22.

[5] Hedner U, Kristensen H, Berntorp E, Ljung R, Petrini P, Tengborn L. Pharmacokinetics of rFVIIa in children. Haemophilia 1998;4:244, Abstract 355.

[6] Villar A, Aronis S, Morfini M, Santagostino E, Auerswald G, Thomsen HF, et al. Pharmacokinetics of activated recombinant coagulation factor VII (NovoSeven®) in children vs. adults with haemophilia A. Haemophilia 2004;10:352—9.

[7] Fridberg MJ, Hedner U, Roberts HR, Erhardtsen E. A study of the pharmacokinetics and safety of recombinant activated factor VII in healthy Caucasian and Japanese subjects. Blood Coagul Fibrinolysis 2005;16:259—66.

[8] Ahlberg Å. Haemophilia in Sweden. VII. Incidence, treatment and prophylaxis of arthropathy and other musculo-skeletal manifestations of haemophilia A and B. Acta Orthopoaedica Scand Suppl 1965;77:3—132.

[9] Hedner U, Ingerslev J. Clinical use of recombinant FVIIa (rFVIIa). Transfus Sci 1998;9:163—76.

[10] Shapiro AD, Gilchrist GS, Hoots WK, Cooper HA, Gastineau DA. Prospective, randomized trial of two doses of rFVIIa (NovoSeven) in haemophilia patients with inhibitors undergoing surgery. Thromb Haemost 1998;80:773—88.

[11] Nilsson IM, Blombäck M, Ahlberg Å. Our experience in Sweden with prophylaxis on haemophilia. In: Proceedings of the Fifth Congress of the World Federation of Haemophilia. Montreal (Quebec, Canada), 1968; Bibl Haemat; 1970;34:111—24; Basel/New York: Karger.

[12] Nilsson IM, Hedner U, Ahlberg Å. Haemophilia prophylaxis in Sweden. Acta Paediatr Scand 1976;65:129—35.

[13] Lindley CM, Sawyer WT, Macik BG, Lusher J, Harrison JF, Baird-Cox K, et al. Pharmacokinetics and pharmacodynamics of recombinant factor VIIa. Clin Pharmacol Ther 1994;55:638−48.

[14] Hedner U. Potential role of recombinant factor VIIa in prophylaxis in severe hemophilia patients with inhibitors. J Thromb Haemost 2006;4:2498−500.

[15] Cooper HA, Jones CP, Campion E, Roberts HR, Hedner U. Rationale for the use of high dose rFVIIa in a high-titre inhibitor patient with haemophilia B during major orthopaedic procedures. Haemophilia 2001;7:517−22.

CHAPTER EIGHT

The Launching and Uses of rFVIIa

Contents

8.1 THE LAUNCHING OF rFVIIa (1996–2000)

At the end of the 1990s there were very few people at Novo Nordisk who believed in rFVIIa. Despite the fact that we had carried out open knee joint surgery on a patient with severe hemophilia without any bleeding complications, using rFVIIa [1,2], they were obviously still not convinced that rFVIIa was of any use or, indeed, if it could be sold at all. I myself was convinced by this time that the addition of rFVIIa could stop and prevent hemorrhages in severe hemophiliacs. I had never experienced that a patient with severe hemophilia could undergo a major operation without the protection of an effective hemostaticum.

One of the greatest problems in the launching of rFVIIa was the lack of knowledge about hemophilia and hemostasis in general at Novo Nordisk's affiliates. They knew a lot about diabetes and insulin but not much about hemophilia. I believe it, therefore, was my personal task to support them at the start and urged them to let me know when they were about to launch in the different countries to make it possible for me to advise them on hemophilia specialists to contact. As I had been active for such a long time in clinical hemophilia care and hemostasis research, I knew most people in these fields, or, at least, I knew who they were.

Treating Life-Threatening Bleedings.
DOI: http://dx.doi.org/10.1016/B978-0-12-812439-0.00008-3

One of the first countries in Europe to launch rFVIIa was France. This took place at a meeting at Montpellier (Fig. 8.1). The title of Novo Nordisk's Satellite Symposium in the official Program was "The Practical

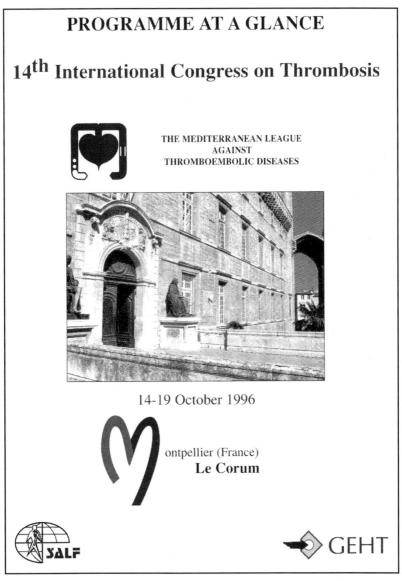

Figure 8.1 The Cover of the Program from the 14th International Congress on Thrombosis, Montpellier, France, Oct 14–19, 1996, where the first launch of rFVII in Europe took place.

Aspects on the Use of NovoSeven®," whereas Novo Nordisk had the title "Symposium on current clinical aspects and future opportunities in the use of NovoSeven® (Recombinant FVIIa)." The interest from the French affiliate was not overwhelming. However, among the other participants in the congress, it was large enough to fill the room allotted to the session where I was to give my lecture. People were even standing along the walls.

For me, it was not really surprising. I knew that the idea of extra rFVIIa as a hemostaticum was unfamiliar to the majority of scientists in this field. I was therefore prepared for an intensive discussion, questioning the whole idea, and did not think so much about the presence or absence of representatives for Novo Nordisk. The discussion turned out to be very lively and constructive. I got the impression that most people got the point in using rFVIIa for the treatment of hemophilia patients with inhibitors.

After the lecture, Jesper Hoiland, then Marketing Manager of Novo Nordisk in France, came up to me and complained about the poor room allotted to the lecture. He was, however, impressed by the auditorium's interest and the lively discussion. He was clearly aware of the great interest in rFVIIa from the audience, as well as its potential importance for hemophilia patients. I had not met Jesper Hoiland previously and thought that, as representative for the commercial part of Novo Nordisk, he might not be so interested in a product that, as yet, had not brought in any money. During our conversation in Montpellier, however, I was very impressed by his ability to see the possibilities of rFVIIa. Based on his impressions from the discussion after my lecture, he was quite convinced that this was something Novo Nordisk should invest in.

Jesper Hoiland's involvement resulted in Novo Nordisk in France investing heavily in rFVIIa and, as far as I know, they were the first affiliate who appointed a special product manager for rFVIIa. This was Jean-Paul Digy, whom I knew from the work on Logiparin, the LMW heparin developed by Novo Nordisk, in the middle of the 1980s. Cooperation with Novo Nordisk in France was extremely successful, and they were, for a long time, the European country that sold most rFVIIa. I must once again stress my admiration for Jesper Hoiland who, without any background in the area of hemostasis, so quickly saw the possibilities for rFVIIa. In my opinion, he was one of the very few at Novo Nordisk who could see this at that time.

The second country to launch rFVIIa was Spain. A symposium with the title "First scientific session on NovoSeven® use in Spain" was held in Cadiz in November 1996 with the aim "to spread the knowledge and clinical experience obtained during registration of the procedure of NovoSeven®." As an example of how information about rFVIIa was spread, even within Novo Nordisk, I can mention the invitation I received from Javier Cortina, Novo Nordisk in Spain, saying that he had attended the World Federation of Hemophilia Congress held earlier in 1996 in Dublin, Ireland, and had been present at the satellite symposium where I presented rFVIIa. He had especially noticed the fact that I had answered questions and participated in the discussion during the entire symposium. I interpreted this as proof of the value of making my experience of hemophilia treatment and my knowledge about rFVIIa, accessible to the affiliates. It strengthened me in my decision to give priority to supporting the different affiliates when they launched the comparatively unknown rFVIIa.

Already in June 1996, the sales of rFVIIa were clearly over the anchor budget (AB) for the year and in January 1997 the month's sales were "way above AB Jan 1995," according to a communication from the Marketing Division on February 2, 1997. For the first 6 months of 1996, England and Sweden were at the top of the list, followed by France in the second place in October 1996. From November 1997 onward, France headed the sales of rFVIIa (almost twice as much as in England) and continued to compete with Germany for first place for several years. The sales of rFVIIa in the following years lay way above the budget and, finally, in 1999—2000 even Novo Nordisk's sales organization was convinced that the product was saleable.

In this connection, I must mention the enormous amount of work done by Pim Tijburg, Novo Nordisk in Holland, as Production Manager for Region 2 that covered France, England, Benelux, Italy, Portugal, Ireland, Spain, Greece, Sweden, and Finland. Pim Tijburg worked hard to spread knowledge about the scientific background of rFVIIa. He used to speak of "scientific marketing." He was also the first one at Novo Nordisk who recognized that rFVIIa in all probability could be used to stop not only hemorrhaging in hemophilia patients but also hemorrhages for other reasons.

Ideas of this kind had arisen at Novo Nordisk in 1997 and were really a follow-up of the patent we had submitted in 1985. This patent covered the use of rFVIIa not only in hemophilia but also to stanch massive hemorrhages in connection with surgery, childbirth, and diffuse bleedings, especially in areas with a high content of proteolytic enzymes, for example, in the gastrointestinal tract and the ear, nose, and throat region. This type of diffuse, massive bleeding is often seen in connection with severe traumata characterized by vast tissue damage. The patent also covered the use of rFVIIa for bleedings in cases of a low platelet count. I shall come back to this later on.

Pim Tijburg started a group he called "Region 2 Skill Team" whose job was, above all, to spread enthusiasm for rFVIIa throughout the company and to convince the Managing Directors at Novo Nordisk of the full potential of rFVIIa. He saw other potential uses of rFVIIa in patients outside the indication of hemophilia, for example, patients with an FVII deficiency, platelet defects (Glanzmann thrombasthenia (GT), thrombocytopenia), and in warfarin-induced bleedings. He worked hard to ensure its further development at Novo Nordisk.

In March 1999, rFVIIa was approved and launched in the United States. It had been available there since 1988, within the so-called "Compassionate Use Program," which meant that the majority of hemophilia specialists already knew about the product and many had personal experience of it. This facilitated the launch.

By this time, the sales of rFVIIa had shown a 40% increase per year since it was first approved in Europe in 1996. At the beginning of 2001, after the launch in the United States and Japan, the sales increased by 63%. rFVIIa apparently filled a need in the global market.

8.2 rFVIIa FOR USE IN AREAS OTHER THAN HEMOPHILIA

8.2.1 rFVIIa and platelet dysfunctions

Already in the beginning of the 1980s, I had the idea that rFVIIa might stop bleedings in conditions other than hemophilia. I thought it is likely that an agent that worked as a hemostatic drug in hemophilia by bypassing FVIII/FIX [3] also should be effective in cases with impaired platelet function. Patients with a low platelet count often have diffuse bleedings especially in areas rich in proteolytic enzyme activity (nosebleeds, mouth and throat bleeding, and gastrointestinal bleeding). Thus, from clinical experience it was clear that these patients form fragile fibrin plugs easily dissolved by proteolytic enzymes. To study a potential effect by rFVIIa in situations characterized by bleedings due to premature dissolution of fibrin plugs by proteolytic enzymes, I used a rabbit model available at the hospital in Malmö. Rabbits with a low platelet count had a prolonged bleeding time following a slight wound inflicted on a small blood vessel. When these rabbits were given a dose of FVIIa, the bleeding time was normalized. I could see, with my own eyes, how proper fibrin plugs were formed in the capillaries at the site of the injury, despite the low number of platelets. The solidity of the fibrin plugs convinced me that FVIIa might be important for hemostasis even under these conditions.

According to the theory about the hemostatic mechanism, current at that time, it was far too remote to suppose that extra FVIIa could influence bleeding caused by too few platelets. In the summer of 1984, I was invited to give a lecture at the so-called Gordon Conference on hemostasis, held once a year in Maine, USA. These conferences were held in high regard by scientists, and only unpublished work was discussed. At the World Federation of Hemophilia Meeting in Stockholm in the summer of 1983, I had given a lecture on the hemostatic effect of pure FVIIa on hemophilia patients with inhibitors. I did not feel that I could talk about the same thing again at the Gordon Conference a year later. Not enough had happened in the field since my last lecture. I decided to talk about my experiments in rabbits with low platelet counts. In the lecture, which was never published according to the rules of the Gordon conferences, I suggested that extra FVIIa might be able to stop bleeding even in patients with a low platelet count. There were no comments and I remember very well Harold Roberts' remark after my presentation that "if it is true, it is fantastic!"

It was clear that no one believed what I said, but the Patent Department at Novo Nordisk was anxious to cover a possible area of use for rFVIIa, and we formulated a patent covering the use of rFVIIa not only for hemophilia but also for bleeding due to premature lysis of fibrin plugs. This might be caused by the formation of loose fibrin plugs (low platelet count) or by an overwhelming concentration of proteolytic enzymes. A high concentration of such enzymes is normally present in special tissues (oral cavity, nose and adjacent tissues, gastrointestinal tract, and urinary tract) or may be increased as a result of extensive tissue damage. Our patent thus included treatment of massive, diffuse hemorrhages in the gastrointestinal tract, the ear, nose, and throat region and hemorrhages in connection with childbirth. The patent was submitted in 1985 and approved as a European patent in 1991. An abstract describing my experiments with rabbits with low platelet counts was published at the American Society of Hematology Meeting in New Orleans on December 1985 [4].

We had tried in the early 1990s to follow up the idea that rFVIIa could be useful in stopping bleeding in patients with a low platelet count. Hemorrhages in these patients are often diffuse and difficult to treat. Characteristics are mucosal bleeding from, for example, the ear, nose, and throat region and from the gastrointestinal tract. The latter are especially difficult to treat. The result of a very limited study of patients with a considerably lowered platelet count was published in connection with the third Novo Nordisk Symposium in Copenhagen in September 1995. Of the eight patients with visible bleeding, two were immediately stanched and in the others the bleeding slowed down. The effect on the bleeding did not entirely correspond to the shortening of the measured bleeding time, and no certain conclusion could be drawn on the effects of rFVIIa on patients with severe thrombocytopenia [5].

A randomized, double-blind, placebo-controlled study of 100 patients with bleeding complications and low platelet counts in connection with a stem-cell transplantation was published in 2005 [6]. The study did not show any significant changes in the bleeding score which was used, but it was, unfortunately, difficult to interpret due to the large variation between patients and the addition of other medication.

During the next years a number of studies were carried out in different models to study the effect of rFVIIa in situations with a lowered platelet count. In our own laboratory in Copenhagen, we studied the effect on the thrombin generation in the cell-based model that was used to illustrate the mechanism of action for rFVIIa in the presence of preactivated platelets. In this model, we could demonstrate that the amount of thrombin, formed in the presence of rFVIIa, was dependent on the number of platelets present, meaning that it decreased in the presence of a lower number of platelets. This confirmed that the effect of rFVIIa is dependent on the presence of platelets. However, we also observed that although the thrombin peak was not completely normalized by rFVIIa, in the presence of subnormal number of platelets, the time taken to initiate thrombin formation after the addition of rFVIIa was shortened [7].

It was previously known that the structure of a formed fibrin plug becomes tighter, the quicker the formation of thrombin takes place [8].

Thus, we drew the conclusion that treating patients with a low platelet count with rFVIIa should contribute to the formation of stronger and more structured fibrin plugs, even if the total thrombin formation was lower than in individuals with a normal number of platelets. Accordingly, the fibrin plugs formed after administration of rFVIIa even in the presence of a low platelet count should be able to stop bleedings more effectively. A later in vitro study of the fibrin structure in blood from patients with a low platelet count, as well as from the one with dysfunctional platelets (GT) showed, with the help of three-dimensional microscope technique, that the loose fibrin structure in both types of patients was normalized after the addition of rFVIIa. A similar pattern was demonstrated in blood from a hemophilia patient stressing the capacity of added rFVIIa to normalize the fibrin structure in patients with defects resulting in a disturbed formation of well-structured fibrin plugs [9].

With the help of flow chamber technique, the Spanish group of Dr. Gines Escolar in Barcelona found that fibrin precipitation in the presence of a damaged part of a vessel wall increased after addition of rFVIIa [10]. Another group in Holland demonstrated that the adhesion of platelets on fibrinogen and tissue-covered glass slides also increased when rFVIIa was added [11]. Over the years, several patient stories have been published relating the positive effect of rFVIIa in patients with a low platelet count. However, no well-controlled clinical study has been carried out.

One group of patients with intractable bleedings are those with functional platelet defects, for example, Glanzmann's thrombasthenia (GT), a congenital defect in a platelet protein (GPIIb/GPIIIa), important for the platelet's ability to adsorb to the vessel wall and to aggregate. Patients with GT are uncommon in Sweden, and I had, therefore, no great experience of treating them, when my colleague from Gothenburg, Lilian Tengborn, in 1996 called me about a little boy, 2 years old, with GT and severe nose bleeding. Despite administering the usual treatment for hemophilia patients with severe nose bleeding, she was unable to stop it. She sounded to be at a loss for what to do, and I remember saying that I did not know what to do either. However, I suggested that she should try rFVIIa, which at the time was approved in Europe for the treatment of hemophilia patients with inhibitors. I also remember that I added that I "have no idea if it will work," but that I had seen a positive effect of rFVIIa on patients with a low platelet count and similar bleedings, and that it "should" work in GT patients. To the relief of us both, it worked very well and I urged her to publish the case, thinking it would be of use to others with similar problems [12].

In the next years, several reports of the successful treatment of severe GT patients and an overview of 59 treated patients (108 bleeding episodes) were published on 1999 and 2004 [13,14]. Based on this material GT patients who, after repeated platelet transfusions, had developed antibodies and therefore no longer had any use of transfusions were approved as an indication for rFVIIa in Europe on 2004 and then in several other countries. It was also considered important to prevent the development of antibodies in young women who, in a possible future pregnancy, ran the risk of transferring these to the fetus, giving rise to complications in the newborn.

Several successful treatments with rFVIIa of bleedings in patients with other functional platelet defects have also been published. One example is the Bernard–Soulier syndrome (defect in glycoprotein Ib, which is important for binding the von Willebrand factor), and another is "platelet storage pool" defects [15].

For all platelet disorders, the same dosage schedule as in hemophilia is recommended, that is, bolus doses of around 270 µg/kg. In especially severe hemorrhages (gastrointestinal hemorrhages, bleeding in the urinary tracts), several doses may be needed.

In this connection, it may also be mentioned that rFVIIa has been successfully used in a number of acquired platelet-related hemorrhages, such as bleeding in kidney diseases and a number of blood diseases where a defect platelet function may develop [15,16; for more references see [15]].

With the increased use of platelet inhibitors to avoid the development of arterial thrombi, for example, in the treatment of myocardial infarction and stroke, the need to be able to counteract the effect of these drugs in connection with acute surgery has increased. Nowadays, many people take drugs that inhibit platelet aggregation by inhibiting GP IIb/IIIa. This gives rise to a condition similar to GT with an accompanying risk for hemorrhaging. rFVIIa has proved useful for treating bleedings resulting from such medication. It has also been found to be useful to prevent bleeding in conjunction with acute surgery during ongoing treatment with platelet inhibitors [17].

Also in patients with von Willebrand's disease, type III and II, a successful use of rFVIIa was reported in 96% of a total of 48 patients with congenital von Willebrand's disease [18]. A patient with mild von

Willebrand's disease with predisposition to allergic problems was successfully treated with rFVIIa in association with an injury bleeding [19].

8.2.2 rFVIIa and liver diseases

Although I had considered the possibility of using rFVIIa for hemorrhages
in patients with impaired liver function, I had never gone any further.
This patient group was, however, included in our first clinical plan from
1988. My first meeting with patients with liver damage, who had actually
been treated with rFVIIa, was at a meeting in Bangkok, Thailand, in the
middle of the 1990s. One of my colleagues had persuaded Novo
Nordisk's representative to arrange a meeting about rFVIIa. She wanted
to present several cases of children with liver diseases whom she had
successfully treated with rFVIIa for bleeding or, to prevent bleeding, in
connection with a liver biopsy.

My colleague was a pediatrician and had spent some months at the
Coagulation Clinic in Malmö in the 1980s. She had heard my name
mentioned there and had, since then, followed my work with rFVIIa. As
soon as it had been licensed for the treatment of hemophilia patients with
inhibitors, she came to the conclusion that it should be useful to stop
bleeding in patients with liver damage, all of whom have an impaired
coagulation mechanism. As many coagulation proteins are produced in
the liver, these will be decreased in the case of poor liver function.
Reduced liver function also leads to increased activity of the enzymes that
dissolve the fibrin plugs and to a low platelet count. Thus, these patients
have a complex disturbance of the hemostatic mechanism and an
increased risk of hemorrhaging.

The results from Bangkok were later presented at the Novo Nordisk
Symposium: "New Aspects on Hemophilia Treatment," in Copenhagen
in May 1999 [20] (Fig. 8.2). There was therefore reason for Novo
Nordisk to consider including liver disease among the possible indications
for use of rFVIIa in addition to hemophilia. In May 1999, the
Nieuwsblad van het Noorden, Holland, published an article about how
the University Hospital in Groningen had used rFVIIa in connection
with five liver transplants on cirrhosis patients. According to the surgeon
in charge, this was the first hospital to show that rFVIIa could be successfully used to decrease bleeding during operations that normally required
many blood transfusions. In the newspaper interview, the same doctor
said that the surgeons, with the help of rFVIIa, had been able to reduce

Figure 8.2 Picture from the Symposium held at Novo Nordisk A/S, Copenhagen, 1999.

the need of blood to three bags instead of the 25 bags normally used in liver transplant surgery. In the presentation given at a symposium at the University Hospital in Groningen the experience was said to be revolutionary and that it would lead to a decreased need for blood transfusions due to massive hemorrhages during surgery. It should also lead to fewer complications in connection with major surgery, as they are known to increase with the number of blood transfusions [21].

Already in 1997, Bernstein et al. had shown that prothrombin time, which is always lengthened in liver disease, was normalized in cirrhosis patients after an injection of rFVIIa [22]. A few years later, it was demonstrated in a flow chamber model containing damaged vascular segments that the addition of rFVIIa to the blood from patients with severely impaired liver function resulted in an improved fibrin formation on the vessel segments [23].

At the previously mentioned Novo Nordisk Symposium in 1999, we had invited several specialists in the field of liver disease, with the aim of obtaining information about potential patient groups who would benefit from rFVIIa, provided we could show that it reduced their bleeding. Finally, clinical studies were initiated in a group of patients who had had a bleeding in the upper gastrointestinal tract (a group that often had renewed bleeding) and a group about to undergo a liver transplant because of severe liver damage.

The first study of six patients with cirrhosis, who underwent a liver transplant, showed that those who were given rFVIIa needed fewer blood transfusions [24]. Unfortunately, these results could not be replicated in two larger studies [25,26]. In the latter of those two studies, it was found, however, that the number of patients who needed no blood transfusion at all was significantly higher in the group treated with rFVIIa.

The reason for these disparate results was discussed in an editorial in *Liver Transplantation*, where it was pointed out that liver transplant technique had vastly improved, with less bleeding as a result. It was also underlined that the role of rFVIIa is perhaps, above all, to stop ongoing bleeding [27]. In consequence of these results, attempts to show a significant effect of rFVIIa in clinical studies were discontinued.

8.2.3 rFVIIa and patients with a normal coagulation system

At the end of the 1990s, when rFVIIa had been approved in Europe, several physicians reported the successful treatment with rFVIIa of various types of patients with bleeding problems, but who had a normal basal coagulation mechanism. These anecdotal case studies covered patients with bleeding in connection with surgery, trauma, and childbirth.

The first publication on a severely bleeding patient without preexisting coagulopathy treated with rFVIIa appeared in 2000 [28], and the first study of rFVIIa given to patients with a basically normal coagulation system was presented at the Novo Nordisk Symposium in 1999. It was an ongoing study of rFVIIa administered during prostate surgery, a procedure known to often be accompanied by severe hemorrhaging. The study was published in its final form in *The Lancet*, 2003, and showed that none of the patients who had been given a dose of 40 µg/kg rFVIIa had needed any form of transfusion and that blood loss decreased significantly [29]. The authors drew the conclusion that a dose of rFVIIa could decrease bleeding and the need for transfusions in connection with major surgery. This study aroused considerable attention, as it also showed that no signs of a general activation of the coagulation mechanism had been observed in patients with normal basal hemostasis.

At the same symposium, I held a lecture with the title "NovoSeven®
as a Universal Hemostatic Agent" [30]. I focused on rFVIIa's mechanism of action, that is, its ability to bind directly to the surface of the activated

platelets and activate FX into FXa, leading to an enhanced thrombin generation. This results in the formation of well-structured fibrin plugs that should be able to stop bleeding more generally.

I remember that I finished my lecture by noting that rFVIIa might be useful for treating massive hemorrhages in connection with surgery, childbirth, bleeding in the gastrointestinal tract and in severe accidents with diffuse, intractable bleeding. The basis for my idea was that these types of hemorrhages often are diffuse with the source of bleeding difficult to localize, and that the increased activity of the enzymes that break down the fibrin plugs and other coagulation proteins, in all probabilities, plays an important role in this type of bleeding. The idea was that rFVIIa, by enhancing the formation of thrombin on the platelet surface, would give rise to stronger fibrin plugs, which would be more resistant toward the attack from proteolytic enzymes. I also expressed my wish to be notified of the results from anyone who used rFVIIa in this type of bleeding. This was, however, heavily criticized by the marketing people at Novo Nordisk because this area of application was outside the licensed indication.

8.2.4 rFVIIa and trauma

In June1999, the first report came of a successfully treated patient in Israel, who had come to the hospital in Tel Aviv with a severe gunshot wound in his stomach. He was already in a bad condition when he arrived at the hospital. It was impossible to keep his blood pressure up, despite blood transfusions through several entries. He was bleeding diffusely and heavily in the entire abdomen. The source of bleeding could not be localized, and it was therefore impossible to stop the bleeding with the help of surgical methods. The doctors gave up and notified his parents that his life could not be saved.

The doctor on call at the coagulation and hemophilia clinic was contacted but did not dare to use rFVIIa as it was not licensed for the treatment of trauma patients. The head of the hemophilia clinic, Dr. Uri Martinowitz, was contacted. He asked them to get hold of all the rFVIIa that was available in the Emergency Clinic and drove at top speed on his motorcycle to the hospital. He got there just as the boy was declared beyond hope of survival. The rFVIIa available was given to the patient, which afterward was found to be a dose of about 60 μg/kg, not an especially high dose in comparison with what is needed to stop hemorrhaging in a hemophilia patient. The bleeding in the Tel Aviv patient did not stop completely but sufficiently for the surgeons to find the damaged vessels and make stitching them up possible. The boy was given a further dose of rFVIIa and survived, which both doctors and relatives regarded as a miracle [31] (Fig. 8.3).

To obtain more support for the use of rFVIIa in patients who had been exposed to severe injuries accompanied by heavy, profuse bleeding, especially when it was difficult to localize the source of bleeding, Uri Martinowitz, together with a few American trauma surgeons, initiated a series of animal experiments where they tried to replicate these situations.

I, myself, participated in the first study. It was carried out at Chaim Sheba Medical Center, Tel HaShomer in Israel, where Uri Martinowitz was the Chief of Israel's National Hemophilia Center. A pig model of a

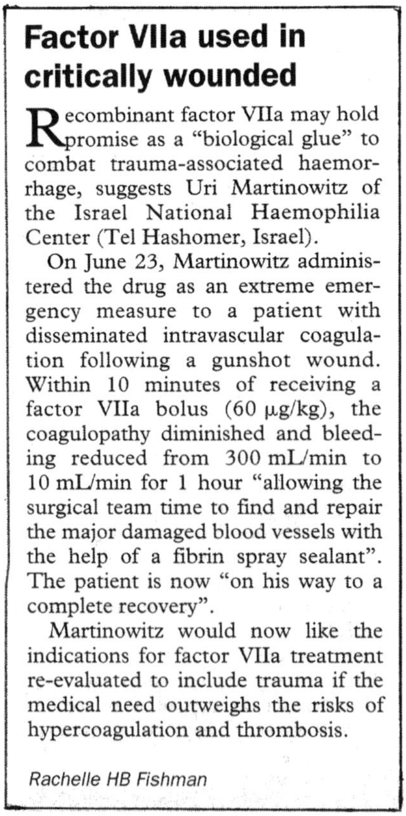

Factor VIIa used in critically wounded

Recombinant factor VIIa may hold promise as a "biological glue" to combat trauma-associated haemorrhage, suggests Uri Martinowitz of the Israel National Haemophilia Center (Tel Hashomer, Israel).

On June 23, Martinowitz administered the drug as an extreme emergency measure to a patient with disseminated intravascular coagulation following a gunshot wound. Within 10 minutes of receiving a factor VIIa bolus (60 µg/kg), the coagulopathy diminished and bleeding reduced from 300 mL/min to 10 mL/min for 1 hour "allowing the surgical team time to find and repair the major damaged blood vessels with the help of a fibrin spray sealant". The patient is now "on his way to a complete recovery".

Martinowitz would now like the indications for factor VIIa treatment re-evaluated to include trauma if the medical need outweighs the risks of hypercoagulation and thrombosis.

Rachelle HB Fishman

Figure 8.3 The report published in *The Lancet* describing the treatment with rFVIIa of the patient heavily bleeding after a gunshot wound.

"grade V liver injury," developed by the American Association for the Surgery of Trauma (AAST), was used. Ten pigs were divided into two groups, one of which was treated with180 μg/kg rFVIIa and the other with saline as a placebo. This model gives a violent hemorrhage that demands intensive treatment with fluids to restore blood pressure. In this first series, we could demonstrate that the amount of FVII in blood sank 52% during the hemorrhage. We also found that the amount of bleeding was 46% less in the pigs who had been given rFVIIa. The results were published in the *Journal of Trauma*, 2001 [32].

It became apparent already during this first study that we had not used the optimal model. The most lively discussions took place between the trauma surgeons who all had extensive experience of trauma surgery in connection with bad traffic accidents, war wounds, and explosive damage after terrorist attacks. It was an interesting and instructive experience for me. I had worked together with surgeons regarding hemophilia patients who also present serious challenges when it comes to finding new, untested solutions in acute situations. I appreciated their refreshing, daring attitudes. Here, in the cramped side room at Sheba Medical Center when more or less wild ideas about how to create a suitable model for studying the effect of rFVIIa on traumatic bleeding were discussed, I was reminded of the atmosphere previously experienced in the hemophilia context.

The problem was how to create a model that would provide a realistic picture of the widespread, profuse bleeding seen as a result of widespread tissue damage caused by a mixture of damage to large and small vessels, as well as muscles. These bleedings are characterized by a local release of all kinds of protein dissolving enzymes. In reality, the treatment cannot be given before the damage has taken place, which must be reflected in the model. No satisfactory solution was found during those intensive days in spring 2000 at Tel Hashomer, but a strong personal network was established. The discussions continued, inspired by the awful injuries in young people subjected to terrorist attacks at this time, during the second "intifada" in Israel, and later by experiences from the wars in Afghanistan and Iraq.

I took part in yet another experimental study where a wound in the aorta from a sharp object was simulated. The initial blood loss was measured and fluid was administered when the blood pressure had decreased substantially. The animals that initially had been given rFVIIa started to rebleed at a significantly higher blood pressure than those that had not received an initial dose of rFVIIa. The initial amount of bleeding was the same for both treated and untreated animals, whereas the bleeding volume after the blood pressure had been raised again was less in the pretreated animals as compared with those that had not received any rFVIIa. The delayed rebleeding in the animals pretreated with rFVIIa suggested that the fibrin plug, formed during the period of low blood pressure, was stronger in the treated animals and thus resisted a higher pressure. We drew the conclusion that rFVIIa had contributed to the formation of a stronger fibrin plug at the site of injury. This study was carried out at the US Army Institute of Surgical Research in Houston, Texas [33].

In connection with heavy blood loss and treatment with blood and other fluids to keep blood pressure at a suitable level marked changes in the coagulation system often developed. This condition is usually called "dilution coagulopathy." At this time, rFVIIa had already been used to stop bleeding in patients with various coagulation disturbances. It was, therefore, a short step to thinking that the addition of rFVIIa to patients with this type of dilution disorder also might decrease bleeding. Good results had been seen in several cases [34−36].

Promptly after the first successful treatment of the patient in Israel, who had an injury through the lower vena cava and diffuse damage to the musculature around the spine, seven critically ill multitransfused trauma patients with coagulation disorders were treated in Israel. Conventional surgical and medical therapy had been ineffective. These seven patients were given rFVIIa in doses between 40 and 120 µg/kg. In all the patients, the bleeding stopped within 5−15 minutes after the administration of 1−3 doses of rFVIIa. The source of bleeding could then be localized and dealt with surgically. At the same time, the need for blood transfusions diminished. Three of the seven patients died as a result of a prolonged period of substantially lowered blood pressure due to heavy bleeding, shock, or blood poisoning with pronounced liver damage. The conclusion was that rFVIIa could be of value used as additional treatment for patients with massive hemorrhages who did not respond to conventional treatment. Controlled studies were recommended to evaluate the effect [37].

At this time, the second "intifada" was going on in Israel, initiating a wave of terrorist attacks at different places in the country. The damage done by homemade bombs filled with nails and other sharp objects, resulted in a new type of injury called "multidimensional" trauma, characterized by extensive damage to several organ systems at the same time.

The result includes explosion damage to the lungs and gastrointestinal tract with torn tissues and widespread bleeding inaccessible for ordinary surgical hemostasis. It also includes damage to arms and legs, which often leads to amputation, in some cases of all the extremities, head wounds, burns, and penetrating wounds in all the inner organs, all of these in the same patient. The cause of death in 90% of these patients is heavy, uncontrollable hemorrhages. The widespread tissue damage gives rise to a complex and intractable coagulation disturbance.

In this situation, rFVIIa proved to be of life-saving importance in several cases. The first report I received about such a patient was in June 2001, when a bomb had been exploded indoors in a discotheque in Tel Aviv. One of the victims was a 15-year-old girl, who was brought to the hospital shortly after the explosion with many wounds caused by metal scrap embedded in the bomb, and causing vast damage to the spine and the large vena cava. She bled profusely and had an extremely low hemoglobin level already on arrival to the hospital. She also had several open fractures. Despite being given a lot of whole blood and different blood components, the hemorrhages continued. After the administration of rFVIIa, the diffuse bleedings quickly dried up and it became possible to surgically repair the vena cava and other damaged vessels. The doctors looking after this patient said it was the worst thing they had ever experienced, although they had participated in several wars and dealt with severe war injuries.

Two other girls, victims of a suicide attack by a car that drove straight into a group of young people who were waiting for a bus one Sunday morning outside of Tel Aviv, had the same type of injuries as the girl above. They also were saved with the help of rFVIIa. I actually spoke to one of these girls 10 years later after she had given birth to her first child. After the accident, she underwent an endless number of operations and it took many years before she could have a more or less normal life. Her mother told me, with delight, that all that remained from her many injuries was a slight limp.

Further victims of terrorist attacks with the same type of injuries were saved with the help of rFVIIa during 2001. In most cases, the massive hemorrhage could be stopped after one or two doses of rFVIIa, which made surgical repair of the multitude of vascular injuries possible. This led, in November 2002, to rFVIIa being approved in Israel, for treating patients with massive bleeding in connection with trauma or surgical procedures. The conditions for use were that the guidelines developed in collaboration with different medical experts were used and that all cases were reported to a register to document the experience.

The results from Israel in the beginning of the 2000s and from the different animal studies carried out from autumn 1999 led to an increased interest for the use of rFVIIa in intractable bleeding episodes in connection with trauma and major surgery. A summary of the animal studies performed in different models was published by John Holcomb in 2005 [38]. These studies had all in common that no thromboembolic side effects were observed in any of them. They varied with regard to the hemostatic effect of rFVIIa when administered alone or in combination with other types of therapy. Two studies showed an increased stability of fibrin plugs and, thus, increased resistance to bleeding. In the same article, Holcomb summarized a number of case studies of patients with severe bleeding, in whom a clear effect of rFVIIa considered as life-saving was observed. The first trauma patient in the United States treated with rFVIIa is given a special mention.

In the autumn 2000, Novo Nordisk worked on designing a clinical study of trauma patients. The first meetings with American trauma surgeons were held in Houston and San Antonio, Texas, in November and December 2000. The difficulties started to pile up at once. It was almost impossible to agree on which patients should be included in the study, not to speak about which dosage should be chosen. All possible and impossible potential risks were discussed.

However, the study was eventually carried out and published in the *Journal of Trauma* in 2005. It comprised two randomized, placebo-controlled double-blind studies. One of the studies included 143 patients who had been subjected to blunt violence and 134 patients with penetrating injury. The patients were randomly allotted to receive either three doses of rFVIIa (200 µg/kg initially, 100 µg/kg an hour later, and 100 µg/kg 3 hours after the first dose) or three doses of placebo. All the patients who had been given rFVIIa showed a decrease in the number of transfusions of red blood cells administered as compared to the placebo group. The decrease was significant in the group that had been subjected to blunt injury. The number of patients needing more than 20 transfusions also decreased in both groups that had received rFVIIa. A higher survival rate was also noted in the patients who were treated with rFVIIa. The number of side effects was evenly distributed between the treatment and the placebo group [39].

It was the intention to follow up this study with another one including more patients to increase the power of the study. It would have a larger impact and was meant to be used for approval of the trauma indication. Much energy and resources were put into the planning of this by Novo Nordisk. An American anesthesiologist, with experience in designing clinical trials and of cooperating with licensing authorities, was employed. The study was, however, discontinued prematurely because the included patients had a lower mortality rate than estimated from previous experience, and the number of patients planned for inclusion, therefore, would not be high enough to show a significant difference between those who had been given rFVIIa and those who had not.

As the study had been carefully planned as a double-blind prospective study and included 150 hospitals in 26 countries, it could teach us a long list of useful lessons. This resulted in several publications. One of them analyzed the importance of the different types of injuries, varying primary care, varying mortality in different countries and emphasized the importance of taking these differences into account when planning large, global studies [40]. Hauser et al. published a report of the entire material in the study, called CONTROL that, when it was stopped, had included 573 patients instead of the planned 1502. It was concluded that:

1. rFVIIa diminished the need for blood products, confirming the result from the previous study;

2. treatment with rFVIIa did not increase the number of the thromboembolic episodes. The number of patients with "acute respiratory distress syndrome" was significantly higher in patients who had been given placebo; and

3. a lower mortality rate was not seen in the treated patients. There was only a tendency to a lower number of patients who developed "multiple organ failure," a dreaded complication in severe traumata.

The study, however, comprised too few patients to show any significant difference, as the group as a whole had a low mortality rate [41]. In the same publication Hauser et al. concluded that a contributing factor to this might be a problem with the inclusion and exclusion criteria in the CONTROL study. Another reason may be that because the study had such a strict protocol regarding general patient care, an overall improvement of general patient care may have occurred and resulted in a decreased mortality. The authors also stressed that as the hemostatic effect of rFVIIa had been proved, efforts to identify those patient groups that would have the greatest benefit from the use of rFVIIa should continue.

I, myself, had no possibility of influencing these activities. My own idea from the beginning had been to focus on a smaller, easily defined patient group, for example, patients with pelvic fractures, which usually give rise to severe hemorrhages and which are rather easy to identify in the acute phase. One of the problems with the treatment of trauma patients is that therapy should be started as early as possible to have maximal effect. I therefore suggested that it would be a good idea to choose a group of patients where a large need of transfusion could reasonably well be foreseen early on. My idea, however, was dismissed at an early stage, as it was considered impossible to extrapolate results from a small group of patients into approval of a general trauma indication.

Another of my suggestions, in this phase of clinical development, was to concentrate on women with massive postpartum hemorrhages due to failed uterus contraction after childbirth. This type of hemorrhage is difficult to treat and is not an uncommon cause of death in connection with childbirth. I had been questioned about the possibility of such a study already at the end of the 1990s by an obstetrician in Great Britain, who offered to contact a sufficient number of centers in Great Britain willing to participate in such a study. A completed study plan was put together with Novo Nordisk in Great Britain but was turned down by the Department of Medical Development in Copenhagen. I had similar suggestions from Great Britain twice more at the end of the 2000s. The reason for continuing to turn down such a study was that the group of patients was too small. The same reason was given later, when similar studies were suggested in developing countries and where it would have been easy to find enough patients for a large study. This type of hemorrhage in connection with delivery is a common cause of mortality in these countries, where the loss of a mother with several other children is a catastrophe for the family.

I had personally experienced how a young woman bled to death because nothing could be done to save her life. This tragic event occurred when I was working as a laboratory technician at the hospital in Malmö, the summer after my first year as a medical student. I was sent to take a blood sample from the patient in question, a young woman who had given birth to her first child. An overwhelming feeling of helplessness characterized the whole situation when the blood, completely lacking any tendency to coagulate, just ran out of her until she died. This dreadful experience came back to me at the beginning of the 1980s. When I saw how hemorrhages in rabbits with a low platelet count (see above) coagulated after the addition of rFVIIa, it convinced me that rFVIIa should be able to stop bleedings in other situations than hemophilia. Typical for the hemorrhages in thrombocytopenic patients and those in women with postpartum bleedings is the formation of unstable fibrin plugs that are easily dissolved by proteolytic enzymes released from disrupted cells at the site of any injury. The result will be the typical widespread, oozing, and profuse bleedings. These bleedings are literally impossible to stop because the exact bleeding source cannot be identified.

As a result of these considerations, the use of rFVIIa in hemorrhages characterized by diffuse bleeding that could not be stopped was included in the patent application from 1985. In the patent, gastrointestinal hemorrhages were given as an example of this type of bleeding.

Unfortunately, the use of rFVIIa in young women suffering from a profuse, life-threatening postpartum bleeding has never been pursued. Thus, profuse postpartum bleeding has never become an approved indication for rFVIIa, which still is a dismal experience for me personally. There are, however, a number of publications that describe the successful treatment with rFVIIa of massive hemorrhages in conjunction with childbirth [37,42−45].

The experiences from Israel, previously described in connection with the second intifada, when several badly injured young people had been saved, lead to an increased interest for rFVIIa in situations other than hemophilia, also in the United States. At this time the United States was involved in the wars in Afghanistan and Iraq. In a number of the *National Geographic Explorer* from April 2003, there was a review of the latest possibilities of saving lives in war situations. Different types of bandages to which had been added different blood substitutes as well as procoagulant substances were developed by the US Army. The same article also mentioned the results obtained with rFVIIa in the treatment of severely injured terrorist victims reported from Israel.

The positive results of the administration of rFVIIa in traumatized patients drew attention to the third factor in the so-called "deadly triad," namely a collapsed coagulation system. The other parts of the "deadly triad" are low body temperature and a disturbed acid−base balance with a decreased pH (acidosis). Since the injured American soldiers were subjected to treatment by the American medical staff quickly, it would be feasible to study the coagulation disorders. Furthermore, different types of blood products were quickly available in the treatment.

These studies showed that the majority of the multiinjured patients had a disturbed coagulation system already by the time they came into care, which suggested immediate treatment with blood products. This approach basically changed trauma treatment and led to the abandonment of intensive fluid treatment with the aim of keeping the blood pressure up, which had been initiated during the Vietnam War. Instead, emphasis is now on the administration of different blood components early in treatment.

This new trauma treatment was presented in a series of publications on collected experiences from different parts of the US Army. It was seen that with the addition of red blood cells, platelet concentrate, and plasma in standardized amounts, the pronounced coagulation problems could almost be avoided. This, in turn, led to a study of the most seriously injured patients, those who had needed ≥ 10 units of red blood cells during 24 hours and who had been treated with rFVIIa. The study showed that patients who had been given rFVIIa before they received red blood cells needed fewer blood transfusions than those who had been given rFVIIa later during the course of treatment. There was no difference between the groups as regards mortality and no increase of thromboembolic side effects [46]. Another study based on the register of trauma patients kept by the US Army since 2003 showed a reduced number of dead after 30 days and no increase of thrombotic side effects [47].

A later study also based on the same register including 2050 patients published in 2010 concluded that rFVIIa tended to be used in the most seriously injured and hemorrhaging patients. This was thought to explain why the patients who had received rFVIIa in this study had been given more blood components and developed more complications than those who had not been given rFVIIa.

This was supported by the fact that no significant difference in mortality nor in frequency of complications was found between the group given rFVIIa and the other group, after a matching system had been used to eliminate the largest deviations regarding the severity of the injuries. The conclusion was drawn that rFVIIa often had been used as a last resort, which was advised against. The publication ends by underlining the need of developing ways to identify trauma patients who would benefit from treatment with rFVIIa [48]. This has, so far, unfortunately not been pursued.

8.2.5 rFVIIa and patients with cerebral hemorrhages

Another completely different use of rFVIIa, which I had not foreseen, was in patients with cerebral hemorrhage. I had imagined that bleeding in the brain, which had already taken place and given rise to symptoms, would not be helped by the addition of rFVIIa. I was, therefore, not very optimistic when one of my coworkers at Novo Nordisk's Marketing Department urged me to contact a neurologist at Columbia University in New York City. My colleague had met him at a meeting in the United States. This neurologist, Dr. Stephan Mayer, had inquired about the possibilities of testing rFVIIa in the treatment of cerebral hemorrhage. As I was in New York City for another reason in the summer 2000, I met Dr. Mayer up and he gave me at least 1 hour's lecture on the latest findings about the progression of a cerebral hemorrhage. The new computed tomography (CT) technique had made it possible to follow the course of a cerebral hemorrhage. Several studies had shown that in 18%–38% of patients, the hemorrhage expanded 2–8 hours after the start. Furthermore, the hematoma growth had been found to be a determinant of mortality and poor outcome after intracerebral hemorrhage. This raised the question of whether treatment results for these patients could be improved by decreasing the volume expansion of the hemorrhage at an early stage.

While I was listening to Stephan Mayer, I came to think about a study by one of the graduate students at the coagulation clinic in Malmö carried out at the beginning of the 1970s. He was a neurosurgeon, who demonstrated that the fibrinolytic activity (the enzymatic activity dissolving fibrin plugs) increased in the spinal fluid in dogs, 1 day after an induced cerebral hemorrhage. He also showed that the blood vessels in the brain were surrounded by tissue, rich in fibrinolytic activity. This led him to speculate on the possibility that the vessel damage leading to a cerebral hemorrhage could lead to increased fibrinolytic activity in the area where the bleeding had taken place. This might disturb the balance between coagulation and the fibrinolytic systems, resulting in the dissolving of formed hemostatic plugs, and thus, may cause the premature dissolution of fibrin plugs intended to stop a hemorrhage.

The picture given by Stephan Mayer, with the expansion of the initial hemorrhage during the next days fitted well with the development of fibrinolysis after an initial cerebral hemorrhage described by Davut Tovi in his dissertation [49]. It also occurred to me that rFVIIa might very well have a good effect in these patients by helping to create strong fibrin plugs at the sites of damage in the small vessels in the brain. These would then become more resistant to premature dissolution by the fibrinolytic activity (Fig. 8.4).

I became convinced that Stephan Mayer's idea was worth trying. However, to do so, Novo Nordisk in both Copenhagen and the United States, had to be convinced that it was a good idea. At this time, Elisabeth Erhardtsen, one of my former coworkers in the hemostasis

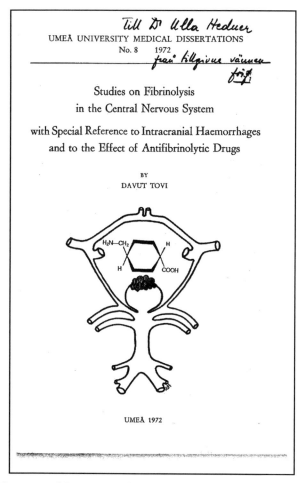

Figure 8.4 The cover of Davut Tovi's dissertation from Umeå University, 1972.

group in Copenhagen, was working at Novo Nordisk in Princeton, USA. She had worked with rFVIIa from the very beginning and was well acquainted with its properties. I told her about my meeting with Stephan Mayer and urged her to visit him and consider his thoughts and suggestions. This eventually resulted in the development of a preliminary study at Novo Nordisk, Princeton. In this study small amounts of rFVIIa was administered, starting on a level of $10-20\,\mu g/kg$, to a limited number of patients at Stephan Mayer's clinic. The study was also discussed with the FDA before it was started.

After having excluded thromboembolic side effects in these first patients, a further larger study was planned to investigate the effect of increasing doses of rFVIIa administered within 3 hours after onset of the hemorrhage. This study in 399 patients compared one placebo group with groups receiving various doses of rFVIIa (40, 80, and $160\,\mu g/kg$). Hemorrhage volume was measured by CT after 24 hours and the clinical picture evaluated after 90 days. Growth in the volume of the intracerebral hemorrhage was significantly decreased in all groups having received rFVIIa in a dose-dependent way as compared with the placebo group. Thromboembolic side effects were seen in 2% of the patients in the placebo group and in 7% of those in the rFVIIa group. Most of the patients had no lasting damage from these events, and the number of fatal or incapacitating complications did not differ significantly between the treated group and those in the placebo group. The mortality after 90 days was 18% compared with 29% in the placebo group. Furthermore, the overall clinical picture was significantly better in the patients who had been treated with rFVIIa [50].

This study was followed up by another one including 841 patients, with 268 in a placebo group. The rest of the patients were divided into two groups receiving rFVIIa in doses of 20 and $80\,\mu g/kg$, respectively. Even in this study, a significant decrease in the hemorrhage volume among the rFVIIa-treated patients was found. However, contrary to the findings in the previous study, the rate of death or severe disability was not reduced. It was stressed by the investigators that more patients with intraventricular hemorrhage at baseline were included in this study. The placebo group contained 29% with this type of bleeding, and the group treated with rFVIIa had 41%. Intraventricular hemorrhage is a well-established determinant of poor outcome after intracerebral hemorrhage. Also, other imbalances between the groups were found. Among the included patients, 82% had an increased blood pressure, 19% had diabetes, 12% had had a previous thromboembolic event, and 5% had acute myocardial infarction [51].

A subanalysis of this study, especially focused on the thromboembolic events, was reported after 2 years [52]. In total 47 venous and 178 arterial events were reported. The venous events did not differ across groups. Of the 178 arterial events, 49 occurred in the placebo group, 47 in the group treated with 20 μg of rFVIIa, and 82 in the 80 μg/kg group. Out of the total 178 arterial events, 141 were acute myocardial events (AMI) and 37 cerebral infarction. Only 38 of the total number of AMIs (141) were reported by the investigators, whereas the majority, 103/141, were detected by the retrospective "Data Monitoring Committee" (DMC) after the study had been finalized. Thus, they had no clinical or any electrocardiographic (EKG) changes indicating an AMI and were identified only by a chemical marker (troponin). Of the 37 cerebral infarction events reported, 9 occurred in the placebo group, 11 in the 20 μg/kg rFVIIa group, and 17 in the 80 μg/kg group. In the 80 μg/kg group, only 8/17 was considered to be possibly related to drug.

Summarizing the two studies by Mayer et al. the effect of rFVIIa in doses of 20 and 40 μg/kg of rFVIIa seemed to decrease the bleeding volume to the same extent, indicating that a dose of 20 μg/kg may be enough to achieve a substantial effect on the bleeding volume. Among the patients who received 20 μg/kg of rFVIIa, the number of arterial side effects was similar to what had been found in the placebo group (27% in the placebo group vs 26% in the group receiving 20 μg/kg of rFVIIa). This may suggest that choosing a dose of 20 μg/kg of rFVIIa would decrease the bleeding volume substantially while not increasing the risk of arterial side effects.

Furthermore, it was concluded that higher doses of rFVIIa, higher age, or the use of platelet aggregation inhibitors (drugs impairing the platelet function and used in patients with an increased risk of developing cardiovascular events), each one on its own, was associated with a higher risk of developing an arterial thromboembolic episode. Diringer et al. [52] suggest that the use of a compulsory biochemical test with the aim of identifying side effects might have resulted in an increased sensitivity to find such. It is also stressed that the troponin biochemical analysis, in fact, often has been found to give falsely increased levels. This may have contributed to explain why three times as many myocardial infarctions were identified by the DMC as compared to those reported by the treating physicians.

The final conclusion of the detailed analysis by Diringer et al. [52] was that higher doses of rFVIIa in a population at high risk for thromboembolic side effects have an insignificantly increased risk for arterial side effects. Thus an increased risk of developing arterial cardiovascular events has not been proved to be associated with rFVIIa treatment in intracranial hemorrhages. Furthermore, in the unlikely case of such an event, it seems to cause minor cardiac damage.

REFERENCES

[1] Hedner U, Ingerslev J. Clinical use of recombinant FVIIa (rFVIIa). Transfus Sci 1998;19:163—76.
[2] Shapiro AD, Gilchrist GS, Hoors WK, Cooper HA, Gastineau DA. Prospective, randomised trial of two doses of rFVIIa (NovoSeven®) in haemophilia patients with inhibitors undergoing surgery. Thromb Haemat 1998;80:773—8.
[3] Hedner U, Kisiel W. Use of human factor VIIa in the treatment of two haemophilia A patients with high-titer inhibitors. J Clin Invest 1983;71:1836—41.
[4] Hedner U, Bergqvist D, Ljungberg J, Nilson B. Haemostatic effect of factor VIIa in thrombocytopenic rabbits. Blood 1985;66(suppl 1): 289a.
[5] Kristensen J, Killander A, Hippe E, Helleberg C, Ellegård J, Holm M, et al. Clinical experience with recombinant factor VIIa in patients with thrombocytopenia. Haemostasis 1996;26(suppl 1):159—64.
[6] Pihusch M, Bacigalupo A, Szer J, von Depka Prondzinski M, Gaspar-Blaudschun B, Hyveled L, et al. Recombinant activated factor VII in treatment of bleeding complications following hematopoietic stem cell transplantation. J Thromb Haemost 2005;3:1935—44.
[7] Kjalke M, Ezban M, Monroe DM, Hoffman M, Roberts HR, Hedner U. High-dose factor VIIa increases initial thrombin generation and mediates faster platelet activation in thrombocytopenia-like conditions in a cell-based model system. Brit J Haemat 2001;114:114—20.
[8] Blombäck B. Fibrinogen and fibrin—proteins with complex roles in haemostasis and thrombosis. Thromb Res 1996;83:1—76.
[9] He S, Jacobsson Ekman G, Hedner U. The effect of platelets on fibrin gel structure formed in the presence of recombinant factor VIIa in haemophilia plasma and in plasma from a patient with Glanzmann thrombasthenia. J Thromb Haemost 2005;3:272—9.
[10] Galan A-M, Tonda R, Pino M, Reverter JC, Ordinas A, Escolar G. Increased local procoagulant action: a mechanism contributing to the favourable hemostatic effect of recombinant FVIIa in plt disorders. Transfusion 2003;43:885—92.
[11] Lisman T, Mosnier LO, Lambert T, Mauser-Bunschoten EP, Meijers JCM, Nieuwenhuis HK, et al. Inhibition of fibrinolysis by recombinant factor VIIa in plasma from patients with severe haemophilia A. Blood 2002;99:175—9.
[12] Tengborn L, Petruson B. A patient with Glanzmann thrombasthenia and epistaxis successfully treated with recombinant factor VIIa. Thromb Haemost 1996;75:981—2.
[13] Poon MC, Demers C, Jobin F, Wu JWY. Recombinant Factor VIIa is effective for bleeding and surgery in patients with Glanzmann thrombasthenia. Blood 1999;94:3951—3.

[14] Poon MC, D'Orion R, von Depka M, Khair K, Négrier C, Karafoulidou A, et al. Prophylactic and therapeutic recombinant factor VIIa administration to patients with Glanzmann's thrombasthenia: results of an international survey. J Thromb Haemost 2004;2:1096—103.

[15] Poon MC. The evidence for the use of recombinant human activated factor VII in the treatment of bleeding patients with quantitative and qualitative platelet disorders. Transfus Med Rev 2007;21:223—36.

[16] Révész T, Arets B, Bierings M. Recombinant FVIIa in severe uremic bleeding. Thromb Haemost 1998;80:204—5.

[17] Altman R, Scazziota A, Herrera MD, Gonzalez C. Recombinant factor VIIa reverses the inhibitory effect of aspirin or aspirin plus clopidogrel on in vitro thrombin generation. J Thromb Haemost 2006;4:2022—7.

[18] Franchini M, Veneri D, Lippi G. The use of recombinant activated factor VII in congenital and acquired von Willebrand's disease. Blood Coagul Fibrinolys 2006;17:615—19.

[19] Alesci S, Krekeler S, Miesbach W. Successful treatment of an injury bleeding on a patient suffering from mild von Willebrand's disease and predisposition to allergic diseases, with recombinant factor VIIa. Haemophilia 2011;17:538—55.

[20] Chuansumrit A, Chantarojanasiri T, Isarangkura P, Teeraratkul S, Hongeng S, Hathirat P. Recombinant activated factor VII in children with acute bleeding resulting from liver failure and disseminated intravascular coagulation. Blood Coagul Fibrinolys 2000;11(suppl 1):101—5.

[21] Nieuwsblad van het Noorden: The Netherlands; Dutch, May 28, 1999.

[22] Bernstein DE, Jeffers L, Erhardtsen E, Reddy KR, Glazer S, Squiban P, et al. Recombinant factor VIIa corrects prothrombin time in cirrhotic patients: a preliminary study. Gastroenterology 1999;113:1930—7.

[23] Tonda R, Galán AM, Pino M, Cirera I, Bosch J, Hernández R, et al. Hemostatic effect of activated recombinant factor VII (rFVIIa) in liver disease: studies in an in vivo model. J Hepatol 2003;39:954—9.

[24] Hendriks HG, Meijer K, de Wolf JTM, Klompmaker IJ, Porte RJ, de Karn PJ, et al. Reduced transfusion requirements by recombinant factor VIIa in orthotopic liver transplantation. Transplant 2001;71:402—5.

[25] Planinsic RM, van der Meer J, Testa G, Grande L, Candela A, Porte RJ, et al. Safety and efficacy of a single bolus administration of recombinant factor VIIa in liver transplantation. Liver Transplant 2005;11:895—900.

[26] Lodge JPA, Jonas S, Jones RM, Olausson M, Mir JP, Soefelt S, et al. Efficacy and safety of repeated perioperative doses of recombinant factor VIIa in liver transplantation. Liver Transplant 2005;11:973—9.

[27] Porte RJ, Caldwell SH. The role of recombinant factor VIIa in liver transplantation. Editorial. Liver Transplant 2005;11:872—4.

[28] Vlot AJ, Ton E, Mackaay AJ, Kramer MHH, Gaillard CAJM. Treatment of a severely bleeding patient without pre-existing coagulopathy with activated recombinant factor VII. Am J Med 2000;108:421—3.

[29] Friederich P, Henny C, Messelink E, Geerdink M, Keller T, Kurth K, et al. Effect of recombinant factor VII on perioperative blood loss in patients undergoing retropubic prostatectomy: a double-blind, placebo-controlled randomised trial. Lancet 2003;361:201—5.

[30] Hedner U. NovoSeven[R] as a universal hemostatic agent. Blood Coag Fibrinolys 2000;11(suppl1):107—11.

[31] Kenet G, Walden R, Eldad A, Martinowitz U. Treatment of traumatic bleeding with recombinant factor VIIa. Lancet 1999;354:1879.

[32] Martinowitz U, Holcomb JB, Pusateri AE, Macaitis JM, Hedner U, Hess JR. Intravenous rFVIIa administered for haemorrhage control in hypothermic coagulopathic swine with grade V liver injuries. J Trauma 2001;50:721−9.

[33] Sondeen JL, Pusateri AE, Hedner U, Yantis LD, Holcomb JB. Recombinant factor VIIa increases the pressure at which rebleeding occurs in porcine uncontrolled aortic haemorrhage model. Shock 2004;22:163−8.

[34] White B, McHale J, Ravi N, Peynolds J, Stephens R, Moriarty J, et al. Successful use of recombinant FVIIa (NovoSeven^R)) in the management of intractable post-surgical intra-abdominal haemorrhage. Brit J Hemat 1999;107:677−8.

[35] Laffan MA, Cummins M. Recombinant factor VIIa for intractable surgical bleeding. Blood 2000;96(suppl 1): 85b.

[36] O'Connell NM, Perry DJ, Hodgson AJ, O'Shaughnessy DF, Laffan MA, Smith OP. Recombinant FVIIa in the management of uncontrolled haemorrhage. Transfusion 2003;43:1711−16.

[37] Martinowitz U, Kenet G, Segal E, Luboshitz J, Lubetskey A, Ingerslev J, et al. Recombinant activated factor VII for adjunctive haemorrhage control in trauma. J Trauma 2001;51:431−8.

[38] Holcomb JB. Use of recombinant activated factor VII to treat the acquired coagulopathy of trauma. J Trauma 2005;58:1298−303.

[39] Boffard KD, Riou B, Warren B, Choong PI, Rizoli S, Roissaint R, et al. Recombinant factor VIIa as adjunctive therapy for bleeding control in severely injured trauma patients: two parallel randomized, placebo-controlled, double-blind clinical trials. J Trauma 2005;59:8−15.

[40] Christensen MC, Parr M, Tortella BJ, Malmgren J, Morris S, Rice T, et al. Global differences in causes, management and survival after severe trauma: the recombinant activated factor VII phase 3 trauma trial. J Trauma 2010;69:344−52.

[41] Hauser CJ, Boffard K, Dutton R, Bernard GR, Croce MA, Holcomb JB, et al. Results of the CONTROL trial efficacy and safety of recombinant activated factor VII in the management of refractory traumatic haemorrhage. J Trauma 2010;69:489−500.

[42] Moscardo F, Perez F, dela Rubia J, Balerdi B, Lorenzo JI, Senent ML, et al. Successful treatment of severe intraabdominal bleeding associated with disseminated intravascular coagulation using recombinant activated factor FVII. Brit J Haemat 2001;114:174−6.

[43] Ahonen J, Jokela R. Recombinant factor VIIa for life-threatening postpartum haemorrhage. Brit J Anaesth 2004;94:592−5.

[44] Karalapillai D, Popham P. Recombinant factor VIIa in massive postpartum haemorrhage. Int J Obstet Anesth 2007;16:29−34.

[45] Bouma LS, Bolte AC, van Geijn HP. Use of recombinant activated factor VII in massive postpartum haemorrhage. Eur J Obstet Gyn Reprod Biol 2008;137:172−7.

[46] Holcomb JB, Jenkins D, Rhee P, Johannigman J, Mahony P, Mehta S, et al. Damage control resuscitation: directly addressing the early coagulopathy of trauma. J Trauma 2007;62:307−10.

[47] Perkins JG, Schreiber MA, Wade JB, Holcomb JB. Early versus late recombinant factor VIIa in combat trauma patients requiring massive transfusion. J Trauma 2007;62:1095−101.

[48] Wade CE, Eastridge BJ, Jones JA, West SA, Sinella PC, Perkins JG, et al. Use of recombinant factor VIIa in US military casualties for a five-year period. J Trauma 2010;69:353−9.

[49] Tovi D. Studies on fibrinolysis in the central nervous system with special reference to intracranial haemorrhages and to the effect of antifibrinolytic drugs. Dissertation at Umeå University, Sweden, 1972.

[50] Mayer SA, Brun NC, Begtrup K, Broderick J, Davis S, Diringer MN, et al. Recombinant activated factor VII for acute intracerebral haemorrhage. N Engl J Med 2005;352:777−85.

[51] Mayer SA, Brun NC, Begtrup K, Broderick J, Davis S, Diringer MN, et al. Efficacy and safety of recombinant activated factor VII for acute intracerebral haemorrhage. N Engl J Med 2008;358:2127−37.

[52] Diringer MN, Skolnick BE, Mayer SA, Steiner T, Davis SM, Brun NC, et al. Thromboembolic events with recombinant activated factor VII in spontaneous intra cerebral haemorrhage. Results from the Factor Seven for Acute hemorrhagic Stroke (FAST) trial. Stroke 2010;41:48−53.

CHAPTER NINE

Mechanism of Action and Dosage

Contents

9.1 MECHANISM OF ACTION

Treatment with extra rFVIIa to stop hemorrhaging was a completely new concept of achieving hemostasis. The role of FVII as the initiator of the clotting process had not been seriously considered previously. Using the cell-based model of hemostasis worked out by the Chapel Hill Group in collaboration with our Novo Nordisk research group in the 1990s, it was demonstrated that pharmacological amounts of rFVIIa bind to the surface of thrombin preactivated platelets with the help of a low-affinity binding site [1−3].

According to current concept normal clotting is initiated by tissue factor (TF), which is normally found in the inner layers of the vessel walls where it is expressed by a number of various cells [4,5]. Following a vessel wall injury, the flexible part of the TF molecule will be exposed to the circulation where it forms a tight complex with FVIIa present in the circulating blood [6,7]. This complex is anchored to the cell membrane through an intramembranous part of the TF molecule [8,9]. TF may also originate from the circulating blood carried by cell elements such as white blood cells, platelets, or microparticles [10−13]. Furthermore, the presence of functional TF-FVIIa complexes extravascularly was demonstrated, which supports a continuous TF-dependent extravascular generation of small amounts of activated FIX and FX [14−16]. In normal individuals, basal concentrations of FX activation peptide and prothrombin activation fragments are present in circulating blood emphasizing the existence of a basal limited continuous activation of the coagulation

Treating Life-Threatening Bleedings.
DOI: http://dx.doi.org/10.1016/B978-0-12-812439-0.00009-5

system extravascularly, which may serve as a "hemostatic sheet" surrounding the vessels [17−19].

The TF-FVIIa-complex activates FX and FV that are present in small quantities on the surface of the damaged cells. The activated FX (FXa) transforms prothrombin to thrombin, and a limited amount of thrombin is formed on the cell surface. This limited amount of thrombin activates platelets and the preactivated platelet surface forms the perfect place for binding FVIII and FIX resulting in the most effective activation of FX and prothrombin [3,20]. The result is the full thrombin burst that transforms the soluble fibrinogen to fibrin. A fibrin plug is formed, which seals the site of injury in the vessel wall. The structure and quality of the fibrin plug is important for its ability to establish a sustainable hemostasis. A well-structured fibrin plug is more resistant to the protein-dissolving enzymes that are released from damaged cells [21−23] (Fig. 9.1).

Patients with hemophilia lack FVIII or FIX, and therefore, the normal generation of thrombin on a preactivated platelet surface does not take place. In these patients, still a limited amount of thrombin is activated with the help of the initial TF-FVIIa complex [20]. This thrombin activates the platelets, but as FVIII or FIX is lacking, full thrombin activation on the platelet surface does not take place. The result is a defect fibrin plug that is loose and porous and easily dissolved by the protein-dissolving enzymes in the environment.

How can extra FVIIa interfere with the generation of a full thrombin burst resulting in the formation of well-structured fibrin plugs resistant to premature lysis? Already in 1987 it was noticed that rFVIIa normalized

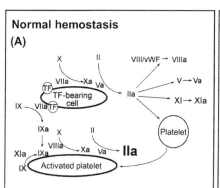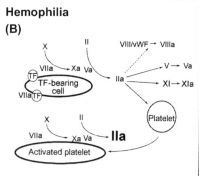

Figure 9.1 Carton of (A) normal hemostasis and (B) hemostasis in hemophilia patients. *Modified from Monroe DM, Hoffman M, Roberts HR. Platelets and thrombin generation. Arterioscler Thromb Vasc Biol 2002;22:1381−9.*

the activated partial thromboplastin time (clotting time in the presence of phospholipids but the absence of TF) suggesting that FVIIa was active also in the absence of TF [24]. This was later confirmed also by others [25−27]. Using the cell-based model, it was demonstrated that rFVIIa binds to preactivated platelets. In the absence of FIX, 10 times higher concentration of rFVIIa was required to generate the same amount of thrombin [1]. Thus rFVIIa binds to preactivated platelets with a low-affinity binding site. Much to my delight I noticed that the amount of rFVIIa required in the absence of FIX (mimicking hemophilia) in the model was similar to the amount of rFVIIa needed for hemostasis in severe hemophilia patients to an extent sufficient to allow major orthopedic surgery to be performed without any bleeding problems [28]. It thus was shown that superphysiological doses of rFVIIa enhances the thrombin generation on activated platelet surface thereby, at least partially, compensating for the lack of FVIII or FIX [29]. A suggested alternative explanation of the therapeutic efficacy of rFVIIa was based on the assumption that the binding of rFVIIa to TF would compete off the FVII-zymogen binding to TF. This would emphasize the TF-dependency of rFVIIa. In the same studies it was, however, demonstrated that platelets substantially increased the hemostatic potential of rFVIIa [30−32].

Thus, the hemostatic effect of rFVIIa is dependent on platelets. However, pharmacological doses of rFVIIa were demonstrated to increase the thrombin generation as well as the platelet adhesion and activation even in situations with low number of platelets [33−36]. Also in other situations characterized by dysfunctional platelets rFVIIa was found to stop bleedings. Thus, successful use was reported in patients with Glanzmann thrombasthenia [37−40].These patients lack functional platelet glycoprotein (GP IIb/IIIa) acting as a receptor of fibrinogen as well as of other adhesive proteins on the platelet surface and rFVIIa was demonstrated to restore the aggregation of platelets lacking this receptor [41]. The resulting enhanced thrombin generation in the presence of rFVIIa in plasma from a patient with Glanzmann's thrombasthenia also was demonstrated to improve the fibrin structure [42]. Thus, the enhanced local thrombin generation mediates the formation of a tightly knit fibrin clot structure, as well as activation of both FXIII and TAFI, all contributing to make the hemostatic fibrin plug more resistant against premature lysis also in patients with low numbers of platelets or such with dysfunctional platelets.

The hemostatic effect of rFVIIa finally depends on its ability to increase thrombin generation at the site of injury resulting in the formation of a well-structured fibrin plug establishing sustainable hemo-stasis at the local site of damage. Furthermore, the increased thrombin generation enhances the activation of the thrombin–activated fibrinolytic inhibitor (TAFI) resulting in an increased resistance toward dissolution of the fibrin plug [43]. An antifibrinolytic effect of rFVIIa dependent on the presence or absence of TAFI also was demonstrated in plasma from hemophilia patients, and its importance for the hemostatic effect of rFVIIa was stressed [44]. The time of sustained hemostasis depends on how long the fibrin plug formed can resist dissolution by the protein-dissolving enzymes. The time is not directly correlated to the amount of FVIIa circulating in the bloodstream [45]. This leads to problems with dosage, which have dominated the development of rFVIIa for clinical use. It is not clear how much extra rFVIIa is needed for a stable and dura-ble hemostasis in a hemophilia patient. At the present time, there is no method measuring the amount of thrombin formed at the site of injury, which thus could indicate the hemostatic effect of the added rFVIIa.

During the last decades, great efforts have been made to develop such methods. Systems have been introduced to measure the formation of thrombin in vitro in the presence of different types of cells. By using whole blood, the presence of both red and white corpuscles and platelets is ensured. However, the necessity of using whole blood requires "bed-side" analyses, which highly limits its usefulness. In other systems, attempts have been made to side step these problems by using either platelet-rich or platelet-poor plasma. If platelet-poor plasma is used, the coagulation system is activated by adding TF or platelet substitute. This makes the system artificial and less useful for determining the effect of rFVIIa or for predicting an effective dose [46–48].

9.2 DOSAGE

Already at the end of the 1990s we showed that rFVIIa disappeared three times more quickly from the bloodstream in children younger than 15 years of age than in adults [49]. This was later confirmed [50]. At the same time, it was also found that platelets from different individuals had varying capacity to generate thrombin on their surface [51]. These obser-vations, together with the findings from the clinical study of rFVIIa in the home treatment of mild to moderate bleedings, that some patients needed several doses of rFVIIa (90 μg/kg per dose) made me aware of the fact that we must take individual variations of the response to rFVIIa into consideration. In children, with a faster clearance rate of rFVIIa from the blood, higher initial doses to achieve an immediate effect would be required. Besides which, one had to be prepared to adjust the dose of rFVIIa upward for individuals who showed a failing initial effect, probably due to a lower capacity to generate thrombin on the surface of their platelets. These findings led to two studies of rFVIIa in a higher dose. I shall come back to this in later chapters.

In the early study, at the University Hospital in Malmö in Spring 1981, of plasma-derived FVIIa (pd-FVIIa) in two hemophilia patients, I had based the dosage on the result of dog studies, which I had carried out, mainly to ensure myself that there were no signs of a general activation of the coagulation system initiated by the addition of pure pd-FVIIa. No such changes were seen after an injection of 50 or 100 U/kg of pd-FVIIa. These doses gave an increase of FVII:C in plasma up to 235% and 500%, respectively. In the treatment of patients, I chose doses of 50 or 100 U/kg pd-FVIIa, which, in these patients, gave FVII concentrations in plasma of 135%−200%. This was almost certainly on the lower side but did not give any signs of a general activation of the coagulation system. For me, the most important thing in these first studies was to be certain that no side effects of the type we had seen in connection with aPCC treatment occurred. The hemostatic effect was however, obvious even after these low doses. However, each patient was given several doses, and patient 1 rebled the next day. Patient 2 was given three doses of pd-FVIIa in conjunction with the loss of a tooth as well as antifibrinolytic treatment locally in the tooth pocket (for details, see Section 2.4).

To reassure myself that the recombinant FVIIa (rFVIIa) was equivalent to pd-FVIIa, I carried out an in vitro test with plasma from a hemophilia patient with inhibitors. rFVIIa or pd-FVIIa were added in increasing amounts to the hemophilia plasma, after which the coagulation time was measured in an APT test ("activated partial thromboplastin test"). A clear shortening of the APT time was achieved after the addition of 1 μg/mL plasma of rFVIIa or pd-FVIIa, whereas a complete normalization of the APT time required the addition of 3.8 μg/mL plasma of rFVIIa or pd-FVIIa. No difference was found between rFVIIa and pd-FVIIa [52]. To achieve a plasma concentration of rFVIIa of 1 μg/mL, a dose of 90 μg/kg would need to be injected (90 μg distributed over 40 mL plasma yields 2.2 μg/mL, and with 47% of injected rFVIIa recovered in the plasma immediately after injection [53], it will end up as approximately 1 μg/mL plasma of rFVIIa). To achieve a plasma concentration of 3.8 μg/mL, which led to complete normalization of the coagulation time in the test tube study, an injected dose of 320 μg/kg would be required. Despite these results indicating that higher doses than those administered in previous studies should be used, lower doses were chosen in the first clinical studies, partly because encouraging results had been achieved in the very first patients who had been given pd-FVIIa in the early 1980s [54].

Already at this time it occurred to me that the fact that FVIIa shortened the APT time did not fit with the claimed idea that FVIIa required TF to be effective. The APT time reagents did not contain TF. They did, however, contain an artificial phospholipid supposed to substitute for platelets.

During the winter of 1988, studies in hemophilia dogs were carried out in collaboration with Dr. Kenneth Brinkhous at Chapel Hill. My first visit to Chapel Hill took place early in 1988, when I discussed the possibility of testing our rFVIIa in the bleeding model developed there. Dr. Brinkhous was at first very skeptical and hesitant to allow his hemophilia dogs to be exposed to rFVIIa. I had to describe in detail the encouraging results of the treatment with pd-FVIIa in the two hemophilia patients published in 1983. Furthermore, I went through why the FVIIa should work as a hemostatic agent based on the various animal models and test tube results it had been through at the time. After some time he finally agreed to a study. A very detailed study protocol was drawn up, and the study was carried out under the personal supervision and active contribution of Dr. Brinkhous.

When he saw the dramatic effect of rFVIIa in doses between 49 and 219 μg/kg in the nail-clipping dog model [55], he became more and more enthusiastic. A dose-related effect, with a moderate shortening of the bleeding time in the dog that had been given 49 μg/kg and a dramatic effect in the dog given 219 μg/kg was achieved. Parallel to this, there was a clear shortening of the partial thromboplastin time (coagulation time) in the dogs. To make sure that the reagents used in the partial thromboplastin—time analysis did not contain any TF, we repeated the test in the presence of an antibody to TF. The partial thromboplastin—time was still shortened confirming the previous results in the APT time but diverging from the generally accepted idea that FVIIa required TF to be active. Not until later in the middle of 1990s I had an opportunity to pursue this question in collaboration with the research group of Harold Roberts also in Chapel Hill.

When it came to decide the dose for the patient about to undergo knee joint surgery under cover of rFVIIa in March 1988, I settled for a dose of 64 μg/kg despite the data in favor of a higher dose (hemophilia dog experiments, test tube studies). This dose was given with an interval of 2−4 hours for the first 24 hours. The patient was also given antifibrinolytic treatment in accordance with the routine used in Scandinavia, in conjunction with surgery in hemophiliacs. He was also operated on by a hand surgeon used to working with complete, painstaking hemostasis during the procedure. We were also several doctors who supervised that dosage, and other treatment was carried out according to all the rules.

At this time, we were not aware of the large individual variation in the capacity for thrombin generation on the preactivated platelet surface. The first signs of an unsatisfactory effect of rFVIIa were not seen until a number of patients had been treated with doses of 70−90 μg/kg. The first patient with unsuccessful rFVIIa treatment in connection with a surgical procedure was in 1989. He was a hemophilia A patient with inhibitors who had developed a massive inguinal hernia (see Section 5.2 for a detailed description). During the operation an initial dose of 72 μg/kg, rFVIIa was prescribed and repeated after 2 hours, after which the same dose was given at longer intervals. Good hemostasis was achieved during the operation, but during the following hours postoperatively the patient started to bleed. His APT time was never normalized [56].

It was also found afterward, that a dose of rFVIIa had been missed, the night after the operation. The patient was put on a continuous infusion of rFVIIa by the local doctor on call. We had already clearly advised against continuous infusion, as we had found that rFVIIa was adsorbed to the walls of the infusion tubing and that, therefore, almost no rFVIIa reached the patient. Eventually, the patient was put on porcine FVIII.

The experience from this second patient undergoing surgery under the cover of rFVIIa led to the next patient with a similar case history and an enormously large inguinal hernia in need of surgery, receiving an initial dose of 90 μg/kg rFVIIa. This patient was operated on in Helsinki and I went there, myself, to ensure that he was given his treatment according to the planned dose schedule. However, this did not help. Foolishly enough, I decreased the second dose to 70 μg/kg that was to be injected every third hour. Excellent hemostasis was noted during the surgical procedure, but during the night after surgery, a dose was missed and, the next morning, bleeding was observed. I immediately increased the dose to 90 μg/kg every third hour. The bleeding stopped and the continued course was without problem [27]. The experience from this operation taught me that higher doses of rFVIIa were necessary to make the treatment less sensitive to deviations in case management.

The dosage in surgical procedures was then changed to 90—120 µg/kg with strict administration every hour for the first 24 hours. With this dosage, full hemostasis could be achieved in 83%—95% of cases [57,58] and in 90%—100% in major surgery including hip and knee joint replacements [28]. The later, controlled study in major surgery confirmed that a dose of 35 µg/kg every hour is not enough to achieve complete hemostasis [59]. Even in the treatment of joint bleedings, doses <90 µg/kg resulted in a suboptimal effect [60].

The conclusion of these early dosing experiences was that I should have trusted my own results in the test tube analyses and those from the dog studies. These studies clearly indicated that higher doses, even up to 300 µg/kg were necessary for full effect of pharmacological doses of rFVIIa. This would perhaps have spared us a great deal of later discussions about dosage and a potential suboptimal effect of rFVIIa, not least internally at Novo Nordisk.

Another lesson learned from these early clinical studies of rFVIIa was that the shorter the time between onset of symptoms and start of treatment the better the effect. This was, however, not a new observation. Everyone dealing with treatment for hemophilia patients knows that the effect of treatment increases markedly, if it is started immediately at the very first indication of an incipient bleeding.

This insight was one of the reasons for introducing home treatment of hemophilia patients at the end of the 1970s. Not least in Sweden, where there often are long distances to health-care facilities, it clearly improved the therapy if the patient had necessary drug available for immediate use at home. A follow-up study, a couple of decades later in Sweden, showed that patients experienced home treatment as the most important improvement in their treatment. It has also been clearly shown that if enough effort is put into teaching the patient and their families to take care of the treatment themselves, at least initially, there will be fewer chronic joint defects.

Already in my first vision of what treatment with rFVIIa should achieve in hemophilia patients with inhibitors, home treatment was an important part. With a certain delay, a home treatment study of rFVIIa was carried out. The patients were given a dose of 90 µg/kg, which they were allowed to repeat three times with a 3-hour interval. If they did not experience any effect after three doses, the treatment was considered a failure.

This study showed a hemostasis efficacy of 92% in a home treatment setting, meaning that the patients administered rFVIIa themselves at home. However, some patients needed more than one dose of 90 µg/kg to stop a mild to moderate bleeding, although treatment was started directly after the onset of symptoms. The aim of home treatment should be that a mild to moderate hemorrhage should be stopped by a single injection if initiated immediately at start of symptoms. Thus, the dose chosen should be high enough to be effective on one single injection. The result of the home treatment study made me immediately aware that many patients needed a higher dose than 90 µg/kg for an optimal effect.

At this time, it had become obvious that there was a considerable variation between the ability of individuals to generate thrombin on the activated surface of platelets. Several studies had also shown that injected rFVIIa disappeared from the bloodstream up to three times faster in children up to 15 years of age, compared to adults [49,50]. It was, thus, obvious that these issues had to be taken into account when choosing doses and the need for a controlled study with higher doses of rFVIIa than were suggested in the approval documents was clear. Work to convince different departments at Novo Nordisk of the importance of such a study get started as soon as possible, already in 1998. The suggestion was to compare the effect of one dose of 270 µg/kg (3 × 90 µg/kg) given as a bolus with the dosage of 3 × 90 µg/kg administered with 3-hour intervals, which had been used in the home treatment study. Above all, it was important to ensure that no side effects appeared in connection with the injection of the higher dose, 270 µg/kg, as a bolus.

The study was discussed for a long time and was given lower and lower priority. After vigorous efforts, however, the studying finally was started in the fall of 2001, about 3 years after the start of discussions. Two studies were decided on—one in the United States and another in Europe. The American study was to include a group of patients receiving FEIBA treatment to allow a comparison between rFVIIa and FEIBA. The European study was completed relatively quickly and published in 2006 [61]. No problems with safety were observed, and the higher dose of 270 µg/kg was approved in Europe and most countries with the exception of the United States.

Despite the fact that the American study took longer time to be completed, the FEIBA arm added to its value. The results from the early 1980s [62,63] with a hemostatic effect of around 60% were confirmed by this American high-dose study published in 2008 [64].

At this time, several studies also had shown that the clearance rate of rFVIIa in children younger than the age of 15 years is up to three times faster than in adults [49,50]. It thus became clear that children younger than 15 years of age should receive higher doses of rFVIIa. Accordingly, a dose of 270 μg/kg is now recommended for children in the countries where this is possible. This higher dose is also recommended in patients showing a less than optimal effect of the dose 90 μg/kg.

REFERENCES

[1] Monroe DM, Hoffman M, Oliver JA, Roberts HR. Platelet activity of high-dose factor VIIa is independent of tissue factor. Brit J Haemat 1997;99:542−7.

[2] Kjalke M, Monroe DM, Hoffman M, Oliver JA, Ezban M, Roberts HR. Active site-inactivated factors VIIa, Xa, and IXa inhibit individual steps in a cell-based model of tissue factor-initiated coagulation. Thromb Haemost 1998;80:578−84.

[3] Monroe DM, Hoffman M, Roberts HR. Platelets and thrombin generation. Arterioscler Thromb Vasc Biol 2002;22:1381−9.

[4] Drake TA, Morrissey JH, Edgington TS. Selective cellular expression of tissue factor in human tissues: Implications for disorders of hemostasis and thrombosis. Am J Pathol 1989;134:1087−97.

[5] Wilcox JN, Smith KM, Schwartz SM, Gordon D. Localization of tissue factor in the normal vessel wall and in the atherosclerotic plaque. Proc Natl Acad Sci USA 1989;86:2839−43.

[6] Wildgoose P, Nemerson Y, Hansen LL, Nielsen FE, Glazer S, Hedner U. Measurement of basal levels of factor VIIa in hemophilia A and B patients. Blood 1992;80:25−8.

[7] Morrissey JH, Macik BG, Neuenschwander PF, Comp PC. Quantitation of activated factor VII levels in plasma using a tissue factor mutant selectively deficient in promoting factor VII activation. Blood 1993;81:734−44.

[8] Edgington TS, Mackman N, Brand K, Ruf W. The structural biology of expression and function of tissue factor. Thromb Haemost 1991;66:67−79.

[9] Rapaport SI, Rao VM. Initiation and regulation of tissue factor-dependent blood coagulation. Arterioscler Thromb 1992;12:1111−21.

[10] Giesen PLA, Rauch U, Bohrmann B, Kling D, Roqué M, Fallon JT, et al. Blood-borne tissue factor: another view of thrombosis. Proc Natl Acad Sci USA 1999;96:2311−15.

[11] Himber J, Wohlgensinger C, RouxS, Damico LA, Fallon JT, Kirchhofer D, et al. Inhibition of tissue factor limits the growth of venous thrombus in the rabbit. J Thromb Haemost 2003;1:889−95.

[12] Lopez-Vilchez I, Escolar G, Diaz-Ricart M, Fuste B, Galan AM, White JG. Tissue factor-enriched vesicles are taken up by platelets and induce platelet aggregation in the presence of factor VIIa. Thromb Haemost 2007;97:202−11.

[13] Rauch U, Bonderman D, Bohrmann B, Badimon JJ, Himber J, Riederer MA, et al. Transfer of tissue factor from leukocytes to platelets is mediated by CD15 and tissue factor. Blood 2000;96:170−5.

[14] Almus FE, Rao LVM, Fleck RA, Rapaport SI. Properties of factor VIIa/tissue factor complexes in an umbilical vein model. Blood 1990;76:354−60.

[15] Hoffman M, Colina CM, McDonald AG, Arepally GM, Pedersen L, Monroe DM. Tissue factor around dermal vessels has bound VII in the absence of injury. J Thromb Haemost 2007;5:1403−8.

[16] Dvorak HN, Senger DR, Dvorak AM, Harvey VS, McDonagh J. Regulation of extravascular coagulation by microvascular permeability. Science 1985;227:1059—61.

[17] Bauer KA, Kass BL, ten Cate H, Bednarek MA, Hawiger JJ, Rosenberg RD. Detection of factor X activation in humans. Blood 1989;74:2007—15.

[18] Bauer KA, Kass BL, ten Cate H, Hawiger JJ, Rosenberg RD. Factor IX is activated in vivo by the tissue factor mechanism. Blood 1990;76:731—6.

[19] Rapaport SI, Rao LVM. The tissue factor pathway: how it has become a "prima Ballerina." Thromb Haemost 1995;74:7—17.

[20] Roberts HR, Monroe DM, White GC. The use of recombinant factor VIIa in the treatment of bleeding disorders. Blood 2004;104:3858—64.

[21] Hedner U, Ezban M. Recent advances in clot biology and assessment of clotting. Trauma 2008;10:261—70.

[22] Collet JP, Lesty C, Montalescot G, Weisel JW. Dynamic changes of fibrin architecture during fibrin formation and intrinsic fibrinolysis of fibrin-rich clots. J Biol Chem 2003;278:21331—5.

[23] Blombäck B, Carlsson K, Fatah K, Hessel B, Procyk R. Fibrin in human plasma: gel architectures governed by rate and nature of fibrinogen activation. Thromb Res 1994;75:521—38.

[24] Hedner U, Lund-Hansen T, Winther D. Comparison of the effect of factor VII prepared from human plasma (pVIIa) and recombinant VIIa (rFVIIa) in vitro and in rabbits. Thromb Haemost 1987;58:270.

[25] Telgt DSC, Macik BG, McCord DM, Monroe DM, Roberts HR. Mechanism by which recombinant factor VIIa shortens the aPTT: activation of factor X in the absence of tissue factor. Thromb Res 1989;56:603—9.

[26] Rao LV, Rapaport SI. Factor VIIa catalyzed activation of factor X independent of tissue factor: its possible significance for control of hemophilic bleeding by infused factor VIIa. Blood 1990;75:1069—73.

[27] Hedner U. Factor VIIa in the treatment of haemophilia. Blood Coagul Fibrinolysis 1990;1:307—17.

[28] Ingerslev J, Friedman D, Gastineau D, Gilchrist G, Johnsson H, Lucas G, et al. Major surgery in haemophilic patients with inhibitors using recombinant factor VIIa. Haemostasis 1996;26(suppl 1):118—23.

[29] Hedner U. Mechanism of action, development and clinical experience of recombinant FVIIa. J Biotechnol 2006;124:747—57.

[30] van't C, Veer NJ, Golden, Mann KG. Inhibition of thrombin generation by the zymogen factor VII: implications for the treatment of hemophilia A by factor VIIa. Blood 2000;95:1330—5.

[31] Butenas S, Brummel KE, Branda RF, Paradis SG, Mann KG. Mechanism of factor VIIa-dependent coagulation in hemophilia blood. Blood 2002;99:923—30.

[32] Butenas S, Brummel KE, Paradis SG, Mann KG. Influence of factor VIIa and phospholipids on coagulation in "acquired" hemophilia. Arterioscler Thromb Vasc Biol 2003;23:123—9.

[33] Kristensen J, Killander A, Hippe E, Helleberg C, Ellegård J, Holm M, et al. Clinical experience with recombinant factor VIIa in patients with thrombocytopenia. Haemostasis 1996;26(suppl 1):159—64.

[34] Vidarsson B, Önundarson P. Recombinant factor VIIa for bleeding in refractory thrombocytopenia. Thromb Haemost 2001;83:634—5.

[35] Kjalke M, Ezban M, Monroe DM, Hoffman M, Roberts HR, Hedner U. High-dose factor VIIa increases initial thrombin generation and mediates faster platelet activation in thrombocytopenia-like conditions in a cell-based model system. Brit J Haematol 2001;114:114—20.

[36] Lisman T, Adelmeijer J, Cauwenberghs S, van Pampus ECM, Heemskerk JWM, de Groot PG. Recombinant factor VIIa enhances platelet adhesion and activation under

flow conditions at normal and reduced platelet count. J Thromb Haemost 2005;3:742−51.

[37] Tengborn L, Petruson B. A patient with Glanzmann thrombasthenia and epistaxis successfully treated with recombinant factor VIIa. Thromb Haemost 1996;75:981−2.

[38] Poon MC, Demers C, Jobin F, Wu JW. Recombinant FVIIa is effective for bleeding and surgery in patients with Glanzmann thrombashenia. Blood 1999;94:3951−3.

[39] Poon MC, D'Oiron R, von Depka M, Khair K, Négrier C, Karafoulidou A, et al. Prophylactic and therapeutic recombinant factor VIIa administration to patients with Glanzmann's thrombasthenia: results of an international survey. J Thromb Haemost 2004;2:1096−103.

[40] Poon MC. Clinical use of recombinant human activated factor VII (rFVIIa) in the prevention and treatment of bleeding episodes in patients with Glanzmann's thrombasthenia. Vasc Health Risk Manag 2007;3:655−64.

[41] Lisman T, Adelmeijer J, Heijnen HF, de Groot PG. Recombinant factor VIIa restores aggregation of αIIbβ3-deficient platelets via tissue factor-independent fibrin generation. Blood 2004;103:1720−7.

[42] He S, Jacobsson Ekman G, Hedner U. The effect of platelets on fibrin gel structure formed in the presence of recombinant factor VIIa in hemophilia plasma and in plasma from a patient with Glanzmann thrombasthenia. J Thromb Haemost 2005;3:272−9.

[43] Mosnier LO, Lisman T, van den Berg HM, Nieuwenhuis HK, Meijers JCM, Bouma BN. The defective down regulation of fibrinolysis in haemophilia A can be restored by increasing the TAFI plasma concentration. Thromb Haemost 2001;86:1035−9.

[44] Lisman T, Mosnier LO, Lambert T, Mauser-Bunschoten EP, Meijers JC, Nieuwenhuis HK, et al. Inhibition of fibrinolysis by recombinant factor VIIa in plasma from patients with severe hemophilia A. Blood 2002;99:175−9.

[45] Cooper HA, Jones CP, Campion E, Roberts HR, Hedner U. Rationale for the use of high dose rFVIIa in a high-titre inhibitor patient with haemophilia B during major orthopaedic procedures. Haemophilia 2001;7:517−22.

[46] Brophy DF, Martin EJ, Christian Barrett J, Nolte ME, Kuhn JG, Gerk PM, et al. Monitoring rFVIIa 90 µg/kg dosing in haemophiliacs: comparing laboratory response using various whole blood assays over 6 h. Haemophilia 2011;17:e949−57.

[47] Kenet G, Stenmo CB, Blemings A, Wegert W, Goudemand J, Krause M, et al. Intra-patient variability of thromboelastographic parameters following in vivo and ex vivo administration of recombinant activated factor VII in haemophilia patients. A multi-centre, randomized trial. Thromb Haemost 2010;103:351−9.

[48] Hemker HC, Giesen P, Al Dieri R, Regnault V, de Smedt E, Wagenvoord R, et al. Calibrated automated thrombin generation measurement in clotting plasma. Pathophysiol Haemost Thromb 2003;33:4−15.

[49] Hedner U, Kristensen HI, Berntorp E, Ljung R, Petrini P, Tengborn L. Pharmacokinetics of rFVIIa in children. Haemophilia 1998;4:3.

[50] Villar A, Aronis S, Morfini M, Santagostino E, Auerswald HF, Thomsen HF, et al. Pharmacokinetics of activated recombinant coagulation factor VII (NovoSeven) in children vs. adults with haemophilia A. Haemophilia 2004;10:352−9.

[51] Sumner WT, Monroe DM, Hoffman M. Variability in platelet procoagulant activity in healthy volunteers. Thromb Res 1996;81:533−43.

[52] Hedner U, Ljungberg J, Lund-Hansen T. Comparison of the effect of plasma-derived and recombinant human FVIIa in vitro and in a rabbit model. Blood Coagul Fibrinolysis 1990;1:145−51.

[53] Lindley CM, Sawyer WT, Macik BG, Lusher J, Harrison JF, Baird-Cox K, et al. Pharmacokinetics and Pharmacodynamics of recombinant factor VIIa. Clin Pharmacol ther 1994;55:638−44.

[54] Hedner U, Kisiel W. The use of human factor VIIa in the treatment of two hemo-philia patients with high-titer inhibitors. J Clin Invest 1983;71:1836—41.

[55] Brinkhous KM, Hedner U, Garris JB, Diness V, Read MS. Effect of recombinant factor VIIa on the hemostatic defect in dogs with hemophilia A, hemophilia B, and von Willebrand disease. Proc Natl Acad Sci USA 1989;86:1382—6.

[56] Gringeri A, Santagostino E, Mannucci PM. Failure of recombinant activated factor VII during surgery in a hemophiliac with high-titer factor VIII antibody. Haemostasis 1991;21:1—4.

[57] Hedner U, Ingerslev J. Clinical use of recombinant FVIIa (rFVIIa). Transfus Sci 1998;19:163—76.

[58] Lusher JM, Ingerslev J, Roberts HR, Hedner U. Clinical experience with recombi-nant factor VIIa. Blood Coagul Fibrinolysis 1998;9:119—28.

[59] Shapiro AD, Gilchrist GS, Hoots WK, Cooper HA, Gastineau DA. Prospective, ran-domized trial of two doses of rFVIIa (NovoSeven) in haemophilia patients with inhi-bitors undergoing surgery. Thromb Haemost 1998;9:119—28.

[60] Lusher JM, Roberts HR, Davignon G, Joist JH, Smith H, Shapiro AD, et al. A randomized double-blind comparison of two dosage levels of recombinant factor VIIa in the treatment of joint, muscle and mucocutaneous haemorrhages in persons with haemophilia A and B, with and without inhibitors. Haemophilia 1998;4: 790—8.

[61] Kavakli K, Makris M, Zulfikar B, Erhardtsen E, Abrams ZS, Kenet G. Home treat-ment of haemarthroses using a single dose regimen of recombinant activated factor VII in patients with haemophilia and inhibitors. A multi-centre, randomized, double-blind, cross-over trial. Thromb Haemost 2006;95:600—5.

[62] Sjamsoedin LJ, Heijnen L, Mauser-Bunschoten EP, van Geijlswijk JL, van Houwelingen H, van Asten P, et al. The effect of activated prothrombin-complex concentrate (FEIBA) on joint and muscle bleeding in patients with hemophilia A and antibodies to factor VIII. A double-blind clinical trial. N Engl J Med 1981;305:717—21.

[63] Lusher JM, Shapiro SS, Palascale JE, Rao AV, Levine PH, Blatt PM. Efficacy of prothrombin-complex concentrates in hemophiliacs with antibodies to factor VIII: a multicenter therapeutic trial. N Engl J Med 1980;303:421—5.

[64] Young G, Shafer FE, Rojas P, Seremetis S. Single 270 microg kg(-1)-dose rFVIIa vs. standard 90 microg kg(-1)-dose rFVIIa and APCC for home treatment of joint bleeds in haemophilia patients with inhibitors: a randomized comparison. Haemophilia 2008;14:287—94.

CHAPTER TEN

The Continued Development of rFVIIa During the 2000s

Contents

In the early and middle of the 2000s, a general dissatisfaction with the development of rFVIIa spread within Novo Nordisk. There was great disappointment that the ambitious studies of rFVIIa in patients with serious traumata and those with cerebral hemorrhages had not yielded conclusive results and, also, the closure of the groups that had specifically worked on these potentially new indications for rFVIIa. As a consequence a considerable number of disappointed people were relocated, mainly to areas outside the field of hemostasis.

One, in all probability, unplanned and unforeseen effect was that the rFVIIa project acquired the general reputation of being a failure. Forgotten was the fact that it had saved many lives, both of hemophilia patients with inhibitors for whom it was primarily developed and also of patients with severe life-threatening hemorrhages of various types such as severe traffic accidents, military violence, bullet wounds, terrorist attacks, difficult, and long drawn-out surgical procedures and hemorrhages in connection with childbirth.

In this situation, I was given the task by our CEO, Lars Rebien Sørensen, of coming up with suggestions about how to improve the continued development. In a document dated March 20, 2003, certain problems were identified with suggestions for changes. I described among other things (1) the lack of scientific competence (without this, I deemed it impossible to plan satisfactory models for further studies), (2) the focus on leadership competence identified as measurable goals for each employee, such as "timelines," number of studies initiated (in the race for

Treating Life-Threatening Bleedings.
DOI: http://dx.doi.org/10.1016/B978-0-12-812439-0.00010-1

"measurable goals," the true goal was forgotten, namely, making rFVIIa available to new patients), and (3) the organization with different people in charge of different clinical studies that did not work satisfactorily. It tended to result in the duplication of work.

The document recommended (1) to gather all competence specialized in hemostasis under the leadership of one person experienced in the field, (2) to focus on quality instead of collecting "measurable results," and (3) to start with small, distinct clinical studies focused on defining the correct dosage in different indications and defining the clinical areas suitable for the use of rFVIIa.

As a follow-up, discussions were held during the following month on how to optimize dosage of rFVIIa by developing methods for monitoring treatment. As part of this a number of improvements and changes were suggested to facilitate the expansion of the indication area of rFVIIa to include all conditions with severe bleeding. It was especially focused on the performance of smaller, well-planned studies directed at finding the optimal dose in different areas of potential use of rFVIIa. On this occasion, Hans Gliese (the then head of the section for "Experimental Medicine") suggested that rFVIIa should be positioned as the best aid in cases of severe hemorrhages and at the same time as a safe alternative to blood products for patients undergoing surgical procedures. His conclusion was that we had a unique product in our hands and that it was up to us to develop it. The expectations in 2003 were regarded as only the top of an iceberg of possible potential. Unfortunately, these suggestions were not realized and shortly afterward Hans Gliese left the company.

The original document of March 2003 stressed, as an important condition for developing new patentable recombinant products, that a production line completely free from the addition of animal serum should be implemented in accordance with the standard for the manufacturing of products based on fermentation of mammalian cells.

10.1 rFVIIa—POTENTIAL LONG-TERM EFFECT

At the beginning of the 2000s, I worked mainly on the two areas that needed rFVIIa use, which I considered required further development to ensure optimal treatment of hemophilia patients with inhibitors. I aimed for a treatment comparable to that offered to patients without inhibitors. A possibility of using higher doses in a home treatment setting should be ensured. Furthermore, I aimed at exploring a potential use of rFVIIa to bring about a long-term decrease in the number of joint bleedings, similar to the effect of so-called prophylaxis with FVIII/FIX concentrate, considered as the golden standard in hemophilia patients without inhibitors.

The conditions for attaining the first aim were created by the two studies comparing a dose of 270 µg/kg administered as one single bolus, with the same amount given in three doses of 90 µg/kg per dose. As described above, these studies were started after a long period of planning in 2003. They were divided into a European study published in 2006 [1], and an American study published in 2008 [2]. Parallel to these, supported by Novo Nordisk, a similar study was carried out in Italy [3]. None of these studies showed any problems with side-effects from the higher dose, and the hemostatic effect was comparable. The approval of the higher dose of rFVIIa and its use in home treatment helped to achieve hemostasis in one single injection (250 µg/kg) also in those patients who required several lower dose injections. Immediate hemostasis is important in hemophilia to prevent accumulation of blood in the joint and subsequent joint damage.

The other issue, which in my opinion needed further exploration, was the signs of a prophylactic effect I observed in a patient who was the first one to be given a high dose of rFVIIa and then underwent orthopedic traction treatment of his knee joint contractures under the cover of rFVIIa. In this patient, long-term comprehensive physiotherapeutic rehabilitation could be carried out without bleeding, under the cover of a daily dose of rFVIIa [4]. Also other cases were reported, in whom a daily dose of rFVIIa was able to break a pattern of constant bleeding and the development of "target joints," marked by highly inflamed joint tissues with repeated bleeding as a result [5,6].

For me, the experiences of the patient with knee joint contractures whom I met during my part-time work at the hemophilia clinic in Malmö at the end of the 1990s (my "key patient #3," Chapter 7: Treatment With rFVIIa in Malmö (1996−99)) were the introduction to closer studies on the distribution and pharmacokinetic of rFVIIa. I had, all along, been aware of the fact that we had not followed up this area. The first pharmacokinetic study, carried out at Chapel Hill in connection with the first clinical dose-effect study of rFVIIa at the end of the1980s, had already shown that injected rFVIIa quickly disappeared from the bloodstream and had a large distribution volume [7]. This indicated that rFVIIa passed through the vessel walls into the extravascular space. It had all along been clear to me that these findings should be followed up. Unfortunately, our research resources during the 1990s had not permitted it.

When, during the latter part of the 1990s, I observed the marked effect of a daily dose of rFVIIa in preventing the breakthrough of bleedings during a time of strenuous physiotherapy in the patient described by Cooper et al. [4], I began to wonder how such an effect could be achieved by a product that disappeared from the bloodstream within 2–3 hours.

Since the 1960s, when hemophilia prophylaxis was started by regular administration of FVIII/FIX concentrate, the aim had been to keep the concentration of coagulation factors in the blood at the level seen in moderate hemophilia. The background for this was that patients with moderate hemophilia had much less severe arthropathy indicating less joint bleeds [8]. It, thus, was concluded that FVIII/FIX plasma levels up to 5% would be enough to prevent the development of severe arthropathy in hemophilia patients. The hemostatic effect was claimed to be correlated to the plasma levels of these factors.

The mechanism of action of rFVIIa is, however, different from the one of FVIII/FIX. When administered in pharmacological doses, rFVIIa is adsorbed to the surface of preactivated platelets. By enhancing the thrombin generation on the platelet surface, it contributes to form a well-structured fibrin plug at the site of vessel wall injury. Thus, its effect is mediated by generating thrombin at the site of injury, which is difficult to measure (for details see Chapter 9: Mechanism of Action and Dosage). The free plasma rFVIIa passes through the vessel walls and distributes in the extravascular tissue. Already in the first patient, in whom I noticed a prolonged clinical effect in terms of prevention of breakthrough bleedings during a long-term intensive physiotherapy, it was obvious that this preventive effect was not directly reflected in the blood concentration of FVII. Despite the fact that the concentration of FVIIa in blood was back to the basic level well before the next rFVIIa injection and, thus, the blood concentration for most of the day was below any level expected to have a hemostatic effect, no breakthrough bleedings occurred [4]. This observation indicated that it was more than only the initial blood concentration that played an important part in the total long-term rFVIIa effect.

To broaden my understanding on this, I found it important to distinguish between the treatment of large hemorrhages, of the type seen in ongoing muscle bleeding, cerebral hemorrhages and bleeding during major surgery, and the more prophylactic long-term effect described in a number of patients where a regular administration of rFVIIa had decreased the number of bleedings [4–6].

In the treatment of fully developed massive hemorrhages, the most important thing seems to be the formation of a strong, well-structured fibrin plug at the site of injury. This fibrin plug must be resistant to premature dissolution by the ubiquitously present protein-dissolving enzymes. The prerequisite for such a fibrin plug to be formed and maintained is a high initial dose of rFVIIa. For maintenance of the hemostatic effect, this dose must be repeated every 2 or 3 hours at least for the first 24 hours [9]. The interval between doses can then be spaced out. This corresponds to the type of FVIII/FIX treatment of patients with hemophilia A/B without inhibitors in similar situations.

The other type of treatment would be the one exploiting the long-term hemostatic effect aiming more at prevention of bleeding. The mechanism of this effect most likely would be to stop microbleedings from microleakages through the vessel wall. Such microbleedings most probably occur in everyone in the normal use of joints and muscles. Availability of coagulation factors in the extravascular space should facilitate the formation of fibrin plugs in the presence of platelets also seeping out through the same microleakages.

From previous studies of the coagulation factors in the lymphatic system, which takes care of fluid from the extravascular space, I knew that almost all coagulation factors are found there [10]. The presence of the coagulation proteins in the extravascular space indicates a close connection between the intravascular and extravascular systems, even under normal circumstances. By studying the relevant literature, I was able to confirm my previous knowledge and found several studies, which, with different methods, demonstrated the presence of coagulation factors in the extravascular space [11].

Through the collaboration between our and Vijay Rao's hemostasis group in Tyler, Texas, I also learned that added rFVIIa was transported into the cells in a laboratory setting and partially broken down. Remaining rFVIIa was returned out to the cell surface. This recycling of rFVIIa continued as long as rFVIIa remained [12]. I drew the conclusion that such a mechanism might play a role in the long-term effect of rFVIIa [13]. From the distribution studies it is known that part of the administered rFVIIa passes through the vessel wall into the extravascular space where cells expressing tissue factor (TF) are found. By forming complexes with TF, rFVIIa may stay on for a longer period of time in the extravascular space and be ready to stop small microbleedings.

Already at the end of the 1990s, I suggested to Novo Nordisk to carry out a controlled study to investigate whether regular administration of rFVIIa to hemophilia patients with inhibitors without any ongoing bleedings would be able to decrease the number of joint bleedings per month. One of the first suggestions, from June 2001, was a study comparing 280 µg/kg doses of rFVIIa three times a week with 75 U/kg FEIBA also administered three times a week. Two years later, in 2003, a study of the regular administration of FEIBA three to four times a week to seven patients with inhibitors was published. In this study, it was concluded that no improvement of already damaged joints could be seen after 3- to 6-year regular administration of FEIBA. However, neither were there any problems with complications [14].

The discussions at Novo Nordisk continued without coming to any agreement over a protocol. Finally, in January 2003, a meeting was held with a number of experienced hemophilia specialists, at which we presented our suggestions and I presented the hypothesis of how rFVIIa might work in this connection. An extremely lively discussion evolved about how a protocol could be designed. The final suggestion involved an observation period of 3 months, during which all bleeding episodes were noted and described in detail. The aim was to confirm that the patients chosen actually had five or more bleeding episodes each month. During the next phase, half of the patients were given a daily dose of 90 µg/kg rFVIIa and the other half a daily dose of 270 µg/kg. This medication was to be continued for 3 months. The patients were then to be followed up for another 3 months. Unfortunately, this study did not get started until April 2004, over a year after the protocol had been drawn up.

When the Novo Nordisk study finally got started, it was carried out in less than 2 years, following the protocol drawn up in January 2003, and was published in 2007 [15]. The results from this controlled study with the aim to elucidate a potential preventive effect of regular administration of rFVIIa to hemophilia patients with inhibitors were positive. In both groups of patients, the number of bleeding episodes decreased significantly during the treatment period, somewhat more in the group given the highest dose. The difference between the groups was not significant, however. It was regarded as surprising at least by me that a daily dose of a product with a half-life of 2.5 hours in the blood had such a marked effect. Even more surprising was that the effect remained during the 3-month follow-up period. The number of bleedings increased somewhat, especially in the group given 90 µg/kg, but in neither group did the number of bleedings return to the original high number of five or more per month.

This effect was primarily explained by the fact that the pattern of repeated bleedings had been broken by the daily administration of rFVIIa and, in consequence, the inflammatory process in the joint tissues had decreased. Later, another research group described a sustained prohemostatic activity of rFVIIa in plasma as well as in platelets in nonbleeding pigs, which they suggested to explain the efficacy of prophylaxis in humans [16]. Because my own experiences of daily administration of rFVIIa in hemophilia patients [4] did not confirm prolonged hemostatic effects in the plasma in terms of FVII concentration and shortening of the prothrombin time (PT), I found it difficult to believe that this was the only explanation for the clinical effect. I wanted to go further and study how injected rFVIIa was distributed in the different body tissues.

The continued discussion at Novo Nordisk after the effect of regular daily administration of rFVIIa had been demonstrated was concentrated to the development of rFVIIa with a prolonged half-life in the blood. All thoughts of a new study of the long-term effects of rFVIIa were rejected. According to my view, the obvious follow-up study of this one, showing a decrease in joint bleedings during and after daily administration of rFVIIa, would have been one trying the administration of rFVIIa two or three times a week instead of a daily dose. At this time I was convinced that this would be enough to ensure an effect. It has later also been proven in local studies from other countries such as Argentina. In Argentina the indication for prophylaxis using rFVIIa three times a week has been approved. However, all suggestions for more studies to obtain a generally approved indication for prophylaxis were rejected by Novo Nordisk. This was motivated by arguments that it would be unacceptably expensive to recommend a daily dose of rFVIIa. Even the observation that, based on clinical experience, it would most probably be enough with a dose two to three times a week was dismissed.

By this time, I had decided to leave hemostasis research at Novo Nordisk's Head Office and, instead, work with the affiliates to spread information about rFVIIa with focus on the importance of using a higher dosage, the importance of home treatment, and the potential possibility of using rFVIIa as prophylaxis to prevent joint bleedings.

The large-scale reorganization of the research department taking place at the time was instrumental in my decision. Adding to my decision was that I had difficulties in supporting some of the new studies and research projects launched at this time. Since the beginning of the 2000s, I had been out of the decision-making group in hemostasis research and I was especially not able to support the new project aiming at constructing an rFVIIa molecule with a longer half-life in the blood.

I was convinced that rFVIIa was effective as a hemostatic product by its ability to contribute to the formation of a well-structured fibrin plug at the site of injury. Accordingly, the capacity to bind to preactivated platelets and to enhance thrombin generation seemed to me to be the most important qualities of the rFVIIa molecule. The new molecules with big carbohydrate molecules linked to the rFVIIa protein to make it stay longer in the blood circulation did not seem to be an attractive way to improve rFVIIa treatment especially because more of the carbohydrate carrying rFVIIa molecule than of rFVIIa was required to generate a similar amount of thrombin both in a reconstituted model system and in a plasma-based model system indicating a lower binding capacity to the preactivated platelet surface [17]. The whole idea with this pegulated rFVIIa (PEG-rFVIIa) molecule was to prevent the diffusion of the rFVIIa through the vessel wall, thus prolonging the time it stays in the plasma. A prolonged hemostatic effect in hemophilia mice was also demonstrated. However, higher concentrations of the PEG-rFVIIa molecule were required to obtain a maximal effect as compared with rFVIIa [18]. Because I, at this time, was convinced about the importance of an extravascular distribution of rFVIIa to make prophylaxis available to hemophilia patients with inhibitors, I did not feel enthusiastic about the new rFVIIa molecule.

Making prophylaxis available for hemophilia patients with inhibitors was, to me, an important step to make their treatment more like the one offered to noninhibitor hemophilia patients. Unfortunately, I could not convince my colleagues in the research department that the new carbohydrate carrying rFVIIa molecule did not fit with the mechanism of action of rFVIIa we had demonstrated 10 years previously. Actually, hemophilia doctors around the world also felt confused because the new ideas put forward by Novo Nordisk did not fit with the previous message emphasizing the generation of thrombin on preactivated platelets, and thus the binding of rFVIIa to the platelets. To me it was obvious that a sustained hemostatic effect of rFVIIa was dependent on the formation of a well-structured fibrin plug at the site of injury being resistant to premature lysis. Furthermore, based on my experience in patients as well as the findings in the trial of prophylaxis use of rFVIIa [15] I was at this time convinced that rFVIIa was distributed extravascularly and that this was important for its long-time effect observed in patients. I found this embarrassing and felt more and more convinced that my time in Novo Nordisk Hemostasis Research was over.

Another problem I had was to explain to hemophilia doctors why Novo Nordisk worked on a subcutaneous formulation of the rFVIIa molecule. In my experience, it is painful to inject a few milliliters of fluid under the skin, whereas an injection in a vein is not especially troublesome for a hemophilia patient after the patient and his parents got used to it. I am also of the decided opinion that the natural immune system should not be provoked by the addition of molecules also only moderated changed as compared to the natural ones. Therefore, the most immunogenic ways of administration should preferably be avoided. Subcutaneous injection is one of these. All such remarks were categorically dismissed and I understood that my time in Novo Nordisk hemostasis research finally was over. I could be of better use elsewhere.

10.1.1 The distribution in the blood of injected rFVIIa

I continued my studies of how rFVIIa is distributed in the body in collaboration with Vijay Rao's group in Tyler, Texas. In the first of our studies, mouse rFVIIa (AF488 fluorophore labeled) was injected into the tail vein of normal mice. The distribution of rFVIIa was monitored using different time intervals. The rFVIIa immediately covered the inner cell layers of the vessel walls. Within an hour, the rFVIIa adsorbed by the vessel walls had moved out to the tissues around the vessels and then spread further to different tissues. It was often localized to areas with TF. The conclusion was drawn that injected rFVIIa is absorbed by the inner cell layers of the vessel walls. It is then transported through the vessel walls into the extravascular space where it most probably binds to TF and remains there for a long time. The prolonged pharmacological effect of rFVIIa seemed to me most likely explained by these findings [19]. In a later additional study, it was demonstrated that the labeled rFVIIa found in the tissues also had biological activity [20]. Furthermore, rFVIIa was seen in certain tissues, for example bone tissue near the growth area of joints, for up to 7 days after the intravenous administration [19].

Other studies have demonstrated that rFVIIa can be absorbed by platelets and transported by them in a maintained active form. This would mean that the effect of rFVIIa can be stored and distributed by platelets [21].

Even if the connection between these findings and a prolonged hemostatic effect of rFVIIa is not completely proven, it could be concluded that prerequisites for such an effect exist.

10.1.2 Qualities of rFVIIa with prolonged half-life in the blood

The modified rFVIIa molecule chosen for the production of rFVIIa with a prolonged half-life in blood was carrying high molecular carbohydrate molecules (PEG-rFVIIa). This molecule proved, as expected, to have approximately twice as long a half-life as rFVIIa in blood [22,23]. In a bleeding model in hemophilia mice, which simulated a massive hemorrhage with damage to both arteries and veins (tail-clipping model), rFVIIa had a better effect 30 minutes after administration, whereas PEG-rFVIIa had a better effect after 6 and 24 hours, which is consistent with a product that takes a longer time to act, but which, then, has the same effect as rFVIIa [18].

However, if my hypothesis of extravascular rFVIIa being responsible for the prolonged hemostatic effect would be correct, the binding to relevant cellular receptors and subsequent internalization of the PEG-rFVIIa as compared to rFVIIa would be of importance. Actually, a study from the Tyler group of Rao et al. found an impaired binding to both endothelial cell protein C receptor and TF on cell surfaces of the PEG-rFVIIa. Furthermore, the internalization of PEG-rFVIIa in endothelial cells and fibroblasts was markedly lower compared to that of rFVIIa [24].

The clinical effect in hemophilia patients with inhibitors was also studied in a randomized, double-blind trial to elucidate the prophylactic effect of PEG-rFVIIa. No dose response was found with a similar number of bleedings in the dose groups (25, 100, and 200 µg/kg). Strangely enough, the number of bleedings decreased already during the last month of the 3-month-long observation time in the patients who had been given the two highest doses. No further effect on the number of bleeding episodes was noted during the treatment period. During the first month of follow-up, the number of bleeding episodes increased markedly and more or less reached the initial values. No correlation was found between the number of bleedings and the levels of FVII concentration in the blood. The number of bleeding episodes rather increased with the higher dose. The possibility that the PEG-rFVIIa molecule with a poorer binding to tissue factor pathway inhibitor (TFPI) results in more free TFPI in the circulation was never mentioned, as far as I can see from the available documents. An increased concentration of TFPI could, at least partially, be the reason for the increased number of bleeding episodes observed in the patients who received higher doses. The conclusion was that the PEG-rFVIIa had no prophylactic effect on bleeding [23].

During spring 2011, the idea of investing in the development of a prophylactic treatment based on rFVIIa was discussed at Novo Nordisk. When I was invited to these discussions, I tried to point out that clinical experience which clearly showed that rFVIIa has a prophylactic effect when administered regularly once or twice a week already existed. It had also clearly been demonstrated that this effect was not clearly correlated to the FVII concentration in the blood. This lack of correlation was also confirmed in the study of the long-term effect of the PEG-rFVIIa. Missing in all this discussion was the mechanism of action of rFVIIa, which differs from that of FVIII and FIX. The whole discussion on prophylaxis in hemophilia was based on the mechanism of action for FVIII/FIX. The functional half-life of injected rFVIIa is dependent on the strength of the fibrin plug generated by the enhanced thrombin generation on the preactivated platelets and not on the concentration of rFVIIa in the blood.

Another important point in the discussion of a possible prophylactic effect of a regular administration of rFVIIa is to keep treatment of an acute, massive hemorrhage, where an ongoing bleeding must be stopped, apart from a treatment directed at stopping small microhemorrhages occurring in the joints during normal physical activity. These latter bleeds are, in all probability, stopped by low concentrations of coagulation factors normally found in the extravascular space. As a proof of ongoing hemostatic activity in the extravascular space, the presence in blood of peptides originating from the activation of coagulation factors has been put forward. These peptides are found in blood from normal individuals [25].

In the treatment of massive ongoing hemorrhages and during major surgery, high doses of rFVIIa must be administered at regular intervals to ensure the formation of stable fibrin plugs at the site of injury. A long-term, more prophylactic effect is probably achieved by rFVIIa in the extravascular space. In all probability, a reservoir of hemostatically active complexes of rFVIIa and TF are formed, ready to form fibrin plugs to stop small microbleedings in the capillaries around the joints. These fibrin plugs most probably prevent the development of larger bleedings. As I understand it, none of this had ever been taken up in connection with future strategy discussions at Novo Nordisk, which I deeply regret.

Currently there is a wide clinical experience of a substantial effect in decreasing the number of joint bleedings by a regular administration of rFVIIa two to three times a week. Therefore, there is no reason to doubt that rFVIIa would provide hemophilia patients with inhibitors a prophylactic effect in terms of fewer joint bleedings similar to what is being achieved in hemophilia patients without inhibitors. In the long term this would hopefully result in a lower risk of developing serious, chronic joint damage also in the patients with inhibitors.

Since the publication of the prophylaxis study using rFVIIa in 2007 [15], three studies using an activated prothrombin complex concentrate (aPCC) have been published [26–28]. They all used dosing at three times a week and found a decrease of bleeding episodes during the prophylaxis period of around 60%. Thus, it is obvious that prophylaxis is feasible also in hemophilia patients with inhibitors by using bypass therapeutic agents. Furthermore, significantly less bleeding episodes occurred in the patients when on prophylaxis as compared with when on demand [27] confirming previous results in noninhibitor patients [29]. It should be emphasized that the aPCC contains a number of blood coagulation proteins including FVII/FVIIa as well as FIX/FIXa, FX/FXa, and other blood proteins such as coagulation inhibitors. Therefore, it should be of no surprise that the administration of aPCC as well as rFVIIa may have a long-term hemostatic effect. Also the aPCC contains procoagulant proteins such as FVIIa, which may distribute into the extravascular compartment thereby enhancing a long-term hemostatic effect. However, it also contains proteins that may counteract hemostasis. Severe bleedings were reported in a few patients included in the prophylaxis study using aPCC as the bypass agent [27]. They all occurred during the washout period between transfer of the patients from prophylaxis to on demand. It may be considered whether an accumulation of counter-hemostatic proteins such as TFPI may have occurred during the prophylaxis period with an administration of the aPCC drug three times a week.

REFERENCES

[1] Kavakli K, Makris M, Zulfikar B, Erhardtsen E, Abrams ZS, Kenet G. Home treatment of haemarthroses using a single dose regimen of recombinant activated factor VII in patients with haemophilia and inhibitors. Thromb Haemost 2006;95:600–5.

[2] Young G, Shafer FE, Rojas P, Seremetis S. Single 270 microg kg(-1)-dose rFVIIa vs. standard 90 microg kg(-1)-dose rFVIIa and APCC for home treatment of joint bleeds in haemophilia patients with inhibitors: a randomized comparison. Haemophilia 2008;14:287—94.

[3] Santagostino E, Mancuso ME, Rocino A, Mancuso G, Scaraggi F, Mannucci PM. A prospective randomized trial of high and standard dosages of recombinant factor VIIa for treatment of hemarthroses in hemophiliacs with inhibitors. J Thromb Haemost 2006;4:367—71.

[4] Cooper HA, Jones CP, Campion E, Roberts HR, Hedner U. Rationale for the use of high dose rFVIIa in a high-titre inhibitor patient with haemophilia B during major orthopaedic procedures. Haemophilia 2001;7:517—22.

[5] Saxon BR, Shanks D, Jory CB, Williams V. Effective prophylaxis with daily recombinant factor VIIa (rFVIIa-NovoSeven) in a child with high titre inhibitors and a target joint. Thromb Haemost 2001;86:1126—7.

[6] Morfini M, Auerswald G, Kobelt RA, Rivolta GF, Rodriguez-Martorell J, Scaraggi FA, et al. Prophylactic treatment of haemophilia patients with inhibitors: clinical experience with recombinant factor VIIa in European haemophilia centres. Haemophilia 2007;13:502—7.

[7] Lindley CM, Sawyer WT, Macik BG, Lusher J, Harrison JF, Baird-Cox K, et al. Pharmacokinetics and pharmacodynamics of recombinant factor VIIa. Clin Pharmcol Ther 1994;55:638—48.

[8] Ahlberg Å. Haemophilia in Sweden VII. Incidence, Treatment and Prophylaxis of Arthropathy and other Musculo-skeletal Manifestations of Haemophilia A and B. Copenhagen: Munksgaard; 1965.

[9] Hedner U, Ingerslev J. Clinical use of recombinant FVIIa (rFVIIa). Transfus Sci 1998;19:163—76.

[10] Le DT, Borgs P, Toneff TW, White MH, Rapaport SI. Hemostatic factors in rabbit limb lymph: relationship to mechanisms regulating extravascular coagulation. Am J Physiol 1998;274:H769—76.

[11] Miller GJ, Howarth DJ, Attfield JC, Cooke CJ, Nanjee MN, Olzewski WI, et al. Hemostatic factors in human peripheral afferent lymph. Thromb Haemost 2000;83:427—32.

[12] Mandal S, Pendurthi UR, Rao LVM. Cellular localization and trafficking of tissue factor in fibroblasts. Blood 2006;107:4746—53.

[13] Hedner U. Potential role of recombinant factor FVIIa in prophylaxis in severe hemophilia patients with inhibitors. J Thromb Haemost 2006;4:2498—500.

[14] Hilgartner MW, Makipernaa A, DiMichele DM. Long-term FEIBA in patients with high-responding inhibitors. Haemophilia 2003;9:261—8.

[15] Konkle BA, Ebbesen LS, Erhardtsen E, Bianco RP, Lissitchkov T, Rusen L, et al. Randomized, prospective clinical trial of recombinant factor VIIa for secondary prophylaxis in hemophilia patients with inhibitors. J Thromb Haemost 2007; 5:1904—13.

[16] Schut AM, Hyseni A, Adelmeijer J, Meijers JCM, de Groot P, Lisman T. Sustained pro-haemostatic activity of rFVIIa in plasma and platelets in non-bleeding pigs may explain the efficacy of a once-daily prophylaxis in humans. Thromb Haemost 2014;112:304—10.

[17] Sørensen BB, Kjalke M, Zopf D, Bjørn SE, Stennicke HR. Platelet-dependent activity of glycoPEGylated rFVIIa. Blood 2007;110:3140.

[18] Holmberg H, Elm T, Karpf D, Tranholm LM, Bjørn SE, Stennicke H, et al. GlycoPEGylated rFVIIa (N7-GP) has a prolonged hemostatic effect in hemophilic mice compared with rFVIIa. J Thromb Haemost 2011;9:1070—2.

[19] Gopalakrishnan R, Hedner U, Ghosh S, Nayak RC, Allen TC, Pendurthi UR, et al. Bio-distribution of pharmacologically administered recombinant factor VIIa (rFVIIa). J Thromb Haemost 2010;8:301–10.

[20] Gopalakrishnan R, Hedner U, Clark C, Pendurthi UR, Rao LVM. rFVIIa transported from the blood stream into tissues is functionally active. J Thromb Haemost 2010;8:2318–21.

[21] Lopez-Vischez I, Tusell J, Hedner U, Altisent C, Escolar G, Galan AM. Traffic of rFVIIa through endothelial cells and redistribution into subendothelium: implications for a prolonged hemostatic effect. J Coag Disorders 2009;1:1–6.

[22] Møss J, Rosholm A, Laurén A. Safety and pharmacokinetics of a glycoPEGylated recombinant activated factor VII derivative: a randomized first human dose trial in healthy subjects. J Thromb Haemost 2011;9:1368–74.

[23] Ljung R, Karim FA, Saxena K, Suzuki T, Arkhammar P, Rosholm A, et al. 40K glycoPEGylated, recombinant FVIIa: 3-month, double-blind, randomized trial of safety, pharmacokinetics and preliminary efficacy in hemophilia patients with inhibitors. J Thromb Haemost 2013;11:1260–8.

[24] Sen P, Ghosh S, Ezban M, Pendurhi UR, Rao LVM. Effect of glycoPEGylation on factor VIIa binding and internalization. Haemophilia 2010;16:339–48.

[25] Bauer KA, Kass BL, ten Cate H, Bednarek MA, Hawiger JJ, Rosenberg RD. Detection of factor X activation in humans. Blood 1989;74:2007–15.

[26] Leissinger CA, Becton DL, Ewing NP, Valentino LA. Prophylactic treatment with activated prothrombin complex concentrate (FEIBA) reduces the frequency of bleeding episodes in paediatric patients with haemophilia A and inhibitors. Haemophilia 2007;13:249–55.

[27] Leissinger C, Gringeri A, Antmen B, Berntorp E, Biasoli C, Charpenter S, et al. Anti-inhibitor coagulant complex prophylaxis in hemophilia with inhibitors. N Engl J Med 2011;365:1684–92.

[28] Antunes SV, Tangada S, Stasyshyn O, Philips J, Guzman-Becerra N, Grigorian A, et al. Randomized comparison of prophylaxis and on-demand regimens with FEIBA NF in the treatment of haemophilia A and B with inhibitors. Haemophilia 2014;20:65–72.

[29] Manco-Johnson MJ, Abshire TC, Shapiro AD, Riske B, Hacker MR, Kilcoyne R, et al. Prophylaxis versus episodic treatment to prevent joint disease in boys with severe hemophilia. N Engl J Med 2007;357:535–44.

CHAPTER ELEVEN

Safety and Health Economy of rFVIIa

Contents

11.1 rFVIIa AND SAFETY

The presence of activated coagulation proteins in the circulation has always been associated with fear of a systemic activation of the coagulation system [1−5]. Thromboembolic (TE) complications in association with the use of activated prothrombin complex concentrates (aPCCs) were also reported. This effect was attributed to activated factors X (Xa) and IX (IXa) [6−8]. This background of reported TE complications associated with the use of aPCC was actually the major reason for me to try to find a better treatment in hemophilia patients with inhibitors. Accordingly, my first focus in the development of rFVIIa was to ensure its safety profile in terms of TE side effects. In fact, the basis for choosing activated FVII (FVIIa) as the attractive candidate for development into a hemostatic drug in hemophilia was the knowledge of the mechanism of action of FVIIa at the time. The basic principle was that FVIIa would not be hemostatically active unless a complex to tissue factor (TF) exposed as a result of an injury to the vessel wall has occurred [9]. The other important issue was the assumption of a low degree of immediate inhibition of an injected FVIIa by circulating coagulation inhibitors such as antithrombin III [10,11], which should make it feasible for injected FVIIa to reach the TF at the site of injury unabated. At the site of injury, hemostasis should be activated leading to the formation of a localized hemostatic plug.

Treating Life-Threatening Bleedings.
DOI: http://dx.doi.org/10.1016/B978-0-12-812439-0.00011-3

To elucidate a potential systemic activation of the coagulation system, animal experiments were performed early in the development of rFVIIa. In studies performed in rabbits, it was demonstrated that doses up to twice the amount later administered to hemophilia patients of recombinant and plasma-derived FVIIa did not induce any laboratory signs of a systemic activation of the coagulation system. There were no clinical symptoms of a generalized shock provoked by the injection of endotoxin and FVIIa in rabbits known to be sensitive to developing intravascular coagulation in the presence of endotoxin [12,13]. Thus pure FVIIa did not seem to induce any generalized activation of the coagulation system in rabbits preinjected by endotoxin. Furthermore, no such symptoms occurred in hemophilia patients later treated with rFVIIa [12].

However, one patient with hemophilia A did develop consumption coagulopathy or disseminated intravascular coagulation (DIC) in association with surgery for a large hip abscess [14,15]. This patient included in the Compassionate Use study in the United States developed the consumption coagulopathy during surgery that focused on removing damaged tissue in the abscess area. The surgery had to be stopped due to the development of a lowered blood pressure and circulatory shock. The consumption reversed, despite continued administration of rFVIIa. The development of consumption coagulopathy in this patient seemed to be more associated with removal of the massive damaged tissue than with the administration of rFVIIa. Another four patients were reported to have disseminated intravascular coagulation (two of them had acquired and two had congenital hemophilia), three had initially been treated with aPPCs. The fourth patient with septic arthritis was reported to have a "subacute DIC" [15].

A review of all spontaneous reports of thrombotic events in patients with congenital or acquired hemophilia with inhibitors reported between the approval of rFVIIa in Europe in 1996 and April 2003 was published in 2004 [15]. At that time, more than 700,000 doses of rFVIIa (90 µg/kg for a 40-kg individual) had been administered in patients with congenital and acquired hemophilia. Relatively few adverse events had occurred. In total 16 thrombotic events, 10 arterial, and 6 venous were spontaneously reported from 1996 through April 2003.

This review from 2004 concludes that many of the cases were quite complicated and could have had multiple factors contributing to the thrombotic event. As a background the incidence of thrombosis within the general population, $1.07-1.17/1000$ in the United States and $1.83/1000$ in France was given. Furthermore, acute myocardial infarction unrelated to replacement therapy has been reported in hemophilia patients and in hemophilia patients receiving aPCC or desmopressin (DDAVP). The authors concluded that it could not be clearly determined in any case that rFVIIa was definitely causally related to the thrombotic event. The review summarizes that the incidence of thrombotic events with the use of rFVIIa was extremely low (less than 1%) and that it appeared to be lower than the one seen with other clotting factor concentrates with known thrombogenic potential [15].

A safety follow-up of reports on TE and fatal events with the use of rFVIIa in congenital and acquired hemophilia between 2003 and 2006, which included approximately 800,000 doses of 90 μg/kg, was published in 2008. A total of 30 TE events and 6 TE-associated fatal events were reported, approximately 3.75 per 100,000 infusions [16]. In this review clinical trials utilizing bolus doses of rFVIIa up to 270 μg/kg were included supporting the absence of safety issues also when using higher doses of rFVIIa, previously reported [17−19]. There were no safety issues reported by using high doses of rFVIIa in a prophylaxis study [20] or in another study of bolus versus continuous infusion in surgery [21].

The report by O'Connell et al. [22] included data collected over around 5.5 years from global reporting of TE events and the use of rFVIIa in both on- and off-label indications. The FDA's Adverse Event Reporting System (AERS) database (MedWatch) was used. The AERS database is a passive surveillance system that receives adverse event reports from product manufacturers, health care professionals, and the public. It is emphasized that the case series does not provide incidence rates for TE adverse events after the use of rFVIIa. The case series usually cannot establish whether the relationship between a product and an event is causal or coincidental, because spontaneous reports have no controls, which are subject to numerous potential reporting biases and have other limitations. Of 168 reports, 151 were from unlabeled indications of rFVIIa and 17 were from the use in hemophilia patients. A history of an underlying medical condition was present in 15% of the cases. It was stressed by the authors that the frequency of TE adverse events cannot be determined from the spontaneous reporting system data likely influenced by numerous factors including time since a product's introduction (Weber effect).

Data from the same program, MedWatch, was used to compare thrombotic event incidence after infusion of rFVIIa versus FVIII inhibitor bypass activity (aPCC, FEIBA) in a study published in 2004 [23]. Also, published case reports were included, and estimated numbers of infusions available from manufacturers were used to assess comparative incidence of thrombotic adverse events in patients receiving rFVIIa or FEIBA. In this material thrombotic events were significantly more frequent in patients who had received rFVIIa than those who had received FEIBA.

The author recognized the limitation of using MedWatch register data as thoroughly described by O'Connell et al. [22]. However, some weaknesses should have been possible to be avoided, for example, the 11 cases in Table 1 of Ref. [23] in the publication being double reported. Furthermore, among the patients listed with adverse events with rFVIIa, several are patients with nonlabeled diagnosis such as patients undergoing liver transplantation, stem-cell transplantation, gastrointestinal bleeding, subarachnoid hemorrhage, a patient with XI deficiency, and others. Several elderly patients with acquired hemophilia having complicated clinical pictures were included as well as patients with thrombophilic disorder. The publication suggests that adverse events of rFVIIa are underreported due to a statement in the package insert of rFVIIa that the risk of TE is "considered to be low." In contrast to this statement, the package insert for rFVIIa did from the beginning include a warning when using rFVIIa in patients with extended atherosclerosis and sepsis because of a tentative risk of an atherosclerotic plaque to rupture at the same time as rFVIIa is injected and thereby facilitate the formation of complexes between rFVIIa and exposed TF at the ruptured plaque [24].

However, no such events have been reported to occur in the post-license follow-up of rFVIIa use. However, it has been emphasized several times that the reason for the low risk of TE events with rFVIIa most probably is the localized effect of the drug. Free rFVIIa in the blood does not activate the coagulation system. It has to be bound to TF exposed at damaged tissue [12,25]. In a Letter to the Editor of *Journal of Thrombosis and Hemostasis*, Sallah et al. also point at the same weaknesses of the publication by Aledort as highlighted above [26].

It should be emphasized that no side effects were observed in healthy volunteers after injection of rFVIIa [27]. Neither were any side-effects observed in a controlled trial including patients without a preformed coagulation disorder undergoing prostatectomy. One single dose of rFVIIa (40 μg/kg) prevented bleedings in these patients [28].

11.1.1 Safety of rFVIIa used off-label

Between 2000 and 2008 the use of rFVIIa in hospitalized patients increased more than 140-fold, and in 2008 ninety seven percent of 18,311 in hospital uses were off-label. During the same time period the inhospital use for hemophilia patients increased less than 4-fold and accounted for 2.7% [29]. In the same publication most of the rFVIIa was reported to be administered in adult and pediatric cardiovascular surgery (29%), body and brain trauma (29%), and intracranial hemorrhage (11%). Across all indications in-hospital mortality was 27%, and 43% of the patients were discharged to home.

It should be emphasized that after a medicine has been through the regulatory process and been approved in the United States by the Food and Drug Administration (FDA), in Europe by the European Medicines Agency (EMA), and by similar Health authorities in Japan and Australia—New Zealand, there are no further limitations on its use [29]. Based on the mechanism of action of rFVIIa, it was obvious that it would have a hemostatic effect also in other situations characterized by profuse, oozing bleedings both in patients with a basically normal hemostasis function and those with various platelet dysfunction. At any site of a vessel wall injury, hemostasis is initiated by the initial formation of TF-FVIIa complexes. These complexes provide the limited amount of thrombin necessary to preactivate platelets. On the surface of these preactivated platelets administered, rFVIIa binds and enhances the full thrombin generation resulting in the formation of tight fibrin hemostatic plugs.

Such well-structured fibrin plugs are necessary for a sustained hemostasis. Because rFVIIa helps to generate tight fibrin plugs at any site of injury, it is obvious that it should be of benefit for hemostasis in all situations with profuse, extended bleedings, for example after big surgery, in trauma, bleedings in connection with deliveries (postpartum bleedings), and extensive gastrointestinal hemorrhages. In most situations characterized by massive tissue damage, various proteolytic enzymes are released. These enzymes degrade any hemostatic plug, especially those who are loose and not well structured. In such situations the administration of rFVIIa will help to restore the capacity of forming tight fibrin plugs resistant to premature degradation, which will be beneficial for hemostasis almost irrespective of the reason for the bleeding [30].

Already during the development of rFVIIa for approval in hemophilia patients with inhibitors, it was used also in other bleeding situations such as platelet dysfunctions, Glanzmann thrombasthenia, thrombocytopenia, other platelet dysfunctions [31,32], bleedings in association with cardiovascular surgery, trauma, liver disease [33,34], intracerebral hemorrhage (ICH) [35—37], and others.

A few deep venous thrombosis have been reported in patients with Glanzmann's thrombasthenia, all having a central venous catheter, which is a well-known trigger for development of thrombosis [38–40].

In a systemic review of patients who had been receiving rFVIIa for off-label indications, the limitations of such analyses are emphasized [41]. Studies available are heterogeneous and too few patients are included, which makes conclusions difficult. Some of the published studies showed a moderate increased risk for TE events. However, it should be stressed that many of the treated patients were very complicated and had many factors predisposing for TE events. No increase in mortality was found across indications for off-label use. Two randomized, controlled studies of patients with ICH showed a dose-dependent increase in arterial thrombosis [36]. A subanalysis of data of the study of 2008 was published later [42] (for details, see Chapter 8: The Launching and Uses of rFVIIa). It may be considered to use a lower dose of rFVIIa in patients with ICH, which still may result in improved results.

In a randomized and controlled study on body trauma patients, no increased risk for TE was demonstrated [43]. The same study showed a reduced need for blood transfusions in the patients treated with rFVIIa resulting in less respiratory (ARDS) complications. However, the study included too few patients for final conclusions regarding efficacy.

11.2 rFVIIa AND HEALTH ECONOMY

The rFVIIa is considered an expensive drug, which has initiated an intensive debate on whether its use would make hemophilia treatment too expensive. It may be worthwhile to emphasize the higher price of all modern drugs requiring a production and high investment cost, for example, new oncology drugs and such for use in inflammatory diseases such as rheumatoid arthritis and others [44,45]. Recently various models for coping with the increased health-care cost have been suggested [45,46].

Severe hemophilia patients without inhibitors who have developed progressive arthropathy are usually subjected to orthopedic surgery, for example, total hip replacement and knee replacement surgery. Such surgery has been contraindicated in patients with inhibitors [47]. In these patients the burden of orthopedic complications and the impact on quality of life have been found to be more severe than in patients without inhibitors [48]. The use of rFVIIa has been established as a successful hemostatic treatment during major orthopedic surgery [49—54].

This will, however, represent a major cost, which has to be balanced against the benefits in terms of reduced pain and suffering, improved function, and mobility. An economic evaluation of major knee surgery in hemophilia patients was performed and published in 2008. It was based on a number of published studies on the effect of major surgery in hemophilia patients with advanced knee arthropathy using rFVIIa for bleeding prophylaxis during the procedure. All patients had moderate or severe pain in the targeted joint before surgery. After surgery 86% of those with data making a comparison feasible, reported no pain at the follow-up. Improvements were also reported regarding functional status. During the first year after the knee surgery, the total cost exceeded the cost in patients with no knee surgery. However, the break-even time between the patients subjected to knee surgery and those without surgery ranged between 6 and 10 years. The authors draw the conclusion that major knee surgery in hemophilia patients with inhibitors treated with rFVIIa might be cost saving in the long term [55].

The cost of using rFVIIa as compared with the plasma-derived aPCC (pd-aPCC) for treating mild to moderate bleedings in hemophilia patients with inhibitors is extensively discussed. The publication by Lyseng-Williamson and Plosker [56] includes a review of 12 different published studies comparing the cost of rFVIIa and that of aPCC in mild to moderate bleedings in inhibitor patients. Despite the acquisition cost of rFVIIa being higher than that of the aPCC, the greater initial efficacy of rFVIIa than of aPCC resulted in lower total medical costs. These experiences stress the importance of comparing total cost of treatment for an episode and not only the cost per vial of drugs used. Achieving hemostasis on one single dose may reduce the cost for additional interventions such as repeated doses, stay in hospital, loss of time at school or work, and rehabilitation cost. In a later study from 2012 a lower total treatment cost per bleed with rFVIIa than with pd-aPCC was confirmed. Time to bleed resolution was also significantly shorter with rFVIIa in this study [57].

REFERENCES

[1] Kasper CK. Postoperative thrombosis in haemophilia B. N Engl J Med 1973;289:160.
[2] Kasper CK. Thromboembolic complications. Thromb Diath Haemorrh 1975;33: 640—4.

[3] Edson JR. Prothrombin-complex concentrates and thrombosis. N Engl J Med 1974;290:403.

[4] Cederbaum AI, Blatt PM, Roberts HR. Intravascular coagulation associated with the use of human prothrombin complex. Ann Intern Med 1976;84:683−7.

[5] White II GC, Roberts HR, Kingdon HS. Prothrombin complex concentrates: potentially thrombogenic materials and clues to the mechanism of thrombosis in vivo. Blood 1977;49:159−70.

[6] Kingdon HS, Lundblad RI, Veltkamp JJ, Aronson DL. Potentially thrombogenic materials in factor IX concentrates. Thromb Diath Haemorrh 1975;33:617−31.

[7] Blatt PM, Lundblad RI, Kingdon HS, McLean G, Roberts HR. Thrombogenic materials in prothrombin complex concentrates. Ann Intern Med 1974;81:766−70.

[8] Hultin M. Activated clotting factors in factor IX concentrates. Blood 1979;54:1028−38.

[9] Prydz H. Studies on proconvertin IV. The adsorption to barium sulphate. Scand J Clin Lab Invest 1964;16:300−6.

[10] Østerud B, Miller-Andersson M, Abildgaard U, Prydz H. The effect of antithrombin III on the activity of the coagulation factors VII, IX and X. Thromb Haemost 1976;35:295−304.

[11] Abildgaard U. Inhibition of thrombin-fibrinogen reaction by antithrombin III, studied by N-terminal analysis. Scand J Clin Lab Invest 1967;20:207−16.

[12] Hedner U. Factor VIIa in the treatment of haemophilia. Blood Coagul Fibrinolysis 1990;1:307−17.

[13] Diness V. rFVIIa in an endotoxin model and in a Wessler model. In: Hedner U, Roberts HR, editors. Proceedings of the 2nd symposium on new aspects of haemophilia treatment. Copenhagen: Medicom Europe; The Netherlands; 1991. p. 147−50.

[14] Stein SF, Duncan A, Cutler DI, Glazer S. DIC in a haemophiliac treated with recombinant factor VIIa. Blood 1990;76:438a.

[15] Abshire T, Kenet G. Recombinant factor VIIa: review of efficacy. Dosing regimens and safety in patients with congenital and acquired factor VIII or IX inhibitors. J Thromb Haemost 2004;2:899−909.

[16] Abshire T, Kenet G. Safety update on the use of recombinant factor VIIa and the treatment of congenital and acquired deficiency of factor VIII or IX with inhibitors. Haemophilia 2008;14:898−902.

[17] Kavakli K, Makris M, Zulfikar B, Erhardtsen E, Abrams ZS, Kenet G. Home treatment of haemarthroses using a single dose regimen of recombinant activated factor VII in patients with haemophilia and inhibitors. A multi-centre, randomized, double-blind, cross-over trial. Thromb Haemost 2006;95:600−5.

[18] Santagostino E, Mancuso ME, Rocino A, Mancuso G, Scaraggi F, Mannuci PM. A prospective randomized trial of high and standard dosages of recombinant factor VIIa for treatment of hemarthroses in hemophiliacs with inhibitors. J Thromb Haemost 2006;4:367−71.

[19] Young G, Shafer FE, Rojas P, Seremetis S. Single 270 microg kg(-1)-dose rFVIIa vs. standard 90 microg kg(-1)-dose rFVIIa and APCC for home treatment of joint bleeds in haemophilia patients with inhibitors: a randomized comparison. Haemophilia 2008;14:287−94.

[20] Konkle BA, Ebbesen LS, Erhardtsen E, Bianco RP, Lissitchkov T, Rusen L, et al. Randomized, prospective clinical trial of recombinant factor VIIa for secondary prophylaxis in hemophilia patients with inhibitors. J Thromb Haemost 2007;5:1904−13.

[21] Pruthi RK, Mathew P, Valentino LA, Sumner MJ, Seremetis S, Hoots WK. Haemostatic efficacy and safety of bolus and continuous infusion of recombinant

factor VIIa are comparable in haemophilia patients with inhibitors undergoing major surgery. Results from an open-label, randomized, multicenter trial. Thromb Haemost 2007;98:726—32.

[22] O'Connell KA, Wood JJ, Wise RP, Lozier JN, Braum MM. Thromboembolic adverse events after use of recombinant human coagulation factor VIIa. JAMA 2006;295:293—8.

[23] Aledort LM. Comparative thrombotic event incidence after infusion of recombinant factor VIIa versus factor VIII inhibitor bypass activity. J Thromb Haemost 2004;2:293—8.

[24] NovoSeven[R] [package insert], Novo Nordisk A/S.

[25] Roberts HR. Recombinant factor VIIa (NovoSeven[R]) and the safety of treatment. Semin Hematol 2001;38(suppl 12):48—50.

[26] Sallah S, Isaksen M, Seremetis S, Payne Rojkjaer L. Comparative thrombotic event incidence after infusion of recombinant factor VIIa vs. factor VIII inhibitor bypass activity—a rebuttal. J Thromb Haemost 2005;3:820—2.

[27] Fridberg MJ, Hedner U, Roberts HR, Erhardtsen E. A study of the pharmacokinetics and safety of recombinant activated factor VII in healthy Caucasian and Japanese subjects. Blood Coagul Fibrinolysis 2005;16:259—66.

[28] Friederich P, Henny C, Messelink E, Geerdink M, Keller T, Kurth K, et al. Effect of recombinant activated factor VII on perioperative blood loss in patients undergoing retropubic prostatectomy: a double-blind, placebo-controlled randomized trial. Lancet 2003;361:201—5.

[29] Logan AC, Yank V, Stafford RS. Off-label use of recombinant factor VIIa in U.S. hospitals: analysis of hospital records. Ann Intern Med 2011;154:516—22.

[30] Hedner U. Mechanism of action, development and clinical experience of recombinant FVIIa. J Biotech 2006;124:747—57.

[31] Tengborn L, Petruson B. A patient with Glanzmann thrombasthenia and epistaxis successfully treated with recombinant factor VIIa. Thromb Haemost 1996;75:981—2.

[32] Poon M-C. The evidence for the use of recombinant human activated factor VII in the treatment of bleeding patients with quantitative and qualitative platelet disorders. Transfus Med Reviews 2007;21:223—36.

[33] Hedner U, Erhardtsen E. Potential role for rFVIIa in transfusion medicine. Transfusion 2002;42:114—24.

[34] Hedner U. Factor VIIa and its potential therapeutic use in bleeding-associated pathologies. Thromb Haemost 2008;100:557—62.

[35] Mayer SA, Brun NC, Begtrup K, Broderick J, Davis S, Diringer MN, et al. Recombinant activated factor VII for acute intracerebral hemorrhage. N Engl J Med 2005;352:777—85.

[36] Mayer SA, Brun NC, Begtrup K, Broderick J, Davis S, Diringer MN, et al. Efficacy and safety of recombinant activated factor VII for acute intracerebral hemorrhage. N Engl J Med 2008;358:2127—37.

[37] Diringer MN, Ferran J-M, Broderick J, Davis S, Mayer SA, Steiner T, et al. Impact of recombinant activated factor VII on health-related quality of life after intracerebral hemorrhage. Cerebrovasc Dis 2007;24:219—25.

[38] Phillips R, Richards M. Venous thrombosis in Glanzmann's thrombasthenia. Haemophilia 2007;13:758—9.

[39] Gruel Y, Pacouret G, Bellucci S, Caen J. Severe Proximal deep vein thrombosis in a Glanzmann thrombasthenia variant successfully treated with a low molecular weight heparin. Blood 1997;90:888—90.

[40] ten Cate H, Brandjes DP, Smits PH, van Mourik JA. The role of platelets in venous thrombosis: a patient with Glanzmann's thrombasthenia and a factor V Leiden mutation suffering from deep venous thrombosis. J Thromb Haemost 2003;1:394—5.

[41] Yank V, Tuohy CV, Logan AC, Bravata DM, Staudenmayer K, Elsenhut R, et al. Systematic review: benefits and harms of in-hospital use of recombinant factor VIIa for off-label indications. Ann Intern Med 2011;154:529−40.

[42] Diringer MN, Skolnick BE, Mayer SA, Steiner T, Davis SM, Brun NC, et al. Risk of thromboembolic events in controlled trials of rFVIIa in spontaneous intracerebral hemorrhage. Stroke 2008;39:850−6.

[43] Boffard KD, Riou B, Warren B, Choong PI, Rizoli S, Roissaint R, et al. Recombinant factor VIIa as adjunctive therapy for bleeding control in severely injured trauma patients: two parallel randomized placebo-controlled, double-blind clinical trials. J Trauma 2005;59:8−15.

[44] Hillner BE, Smith TJ. Efficacy does not necessarily translate to cost effectiveness: a case study in the challenges associated with 21st-century cancer drug pricing. J Clin Oncol 2009;27:2111−13.

[45] Messori A, Trippoli S, Innocenti M, Morfini M. Risk-sharing approach for managing factor VIIa reimbursement in haemophilia patients with inhibitors. Haemophilia 2010;16:545−66.

[46] van Doorslaer E, O'Donnell O, Ranuan-Elyia R, Somanathan A, Raj Adhikari S, Gang C, et al. Effects of payments for health care on poverty estimates in 11 countries in Asia: an analysis of household survey data. The Lancet 2006;368:1357−64.

[47] Hedner U, Lee CA. First 20 years with recombinant FVIIa (NovoSeven). Haemophilia 2011;17:e172−82.

[48] Morfini M, Haya S, Tagariello G, Pollmann H, Quintana M, Siegmund B, et al. European study on orthopaedic status of haemophilia patients with inhibitors. Haemophilia 2007;13:606−12.

[49] Shapiro AD, Gilchrist GS, Hoots KW, Cooper HA, Gastineau DA. Prospective, randomized trial of two doses of rFVIIa (NovoSeven) in haemophilia patients with inhibitors undergoing surgery. Thromb Haemost 1998;80:773−8.

[50] Ingerslev J. Efficacy and safety of recombinant factor VIIa in the prophylaxis of bleeding in various surgical procedures in hemophilic patients with factor VIII and factor IX inhibitors. Semin Thromb Hemost 2000;26:0425−32.

[51] Hvid I, Rodriguez-Merchan EC. Orthopaedic surgery in haemophilic patients with inhibitors: an overview. Haemophilia 2002;8:288−91.

[52] Rodriguez-Merchan EC, Wiedel JD, Wallny T, Hvid I, Berntorp E, Rivard G-E, et al. Elective orthopaedic surgery for inhibitor patients. Haemophilia 2003;9:625−31.

[53] Rodriguez-Merchan EC, Quintana M, Jimenez-Yuste V, Hernandez-Navarro F. Orthopaedic surgery for inhibitor patients: a series of 27 procedures (25 patients). Haemophilia 2007;13:613−19.

[54] Caviglia H, Candela M, Galatro G, Neme D, Moretti N, Bianco RP. Elective orthopaedic surgery for haemophilia patients with inhibitors single centre experience of 40 procedures and review of the literature. Haemophilia 2011;17:910−19.

[55] Ballal RD, Botteman MF, Foley I, Stephens JM, Wilke CT, Joshi AV. Economic evaluation of major knee surgery with recombinant activated factor VII in haemophilia patients with high titer inhibitors and advanced knee arthropathy: exploratory results via literature-based modelling. Current Medical Research and Opinions 2008;24:753−68.

[56] Lyseng-Williamson KA, Plosker GL. Recombinant factor VIIa (eptacog alfa): a pharmacoeconomic review of its use in haemophilia in patients with inhibitors to clotting factors VIII or IX. Pharmacoeconomics 2007;25:1007−29.

[57] Salaj P, Penka M, Smejkal P, Geierova V, Ovesná P, Brabec P, et al. Economic analysis of recombinant activated factor VII versus plasma-derived activated prothrombin complex concentrate in mild to moderate bleeds: haemophilia registry data from the Czech Republic. Thromb Res 2012;129:e233−7.

CHAPTER TWELVE

Three Different Descriptions of How rFVIIa Was Developed

Contents

I have given my version of the background and origin of rFVIIa as a treatment of life-threatening bleedings and will now try to compare it with two earlier descriptions and comment on them. The first one was written at the beginning of the 2000s for distribution within the company and has the title *The NovoSeven®️ Story* (Novo Nordisk 2004, Photos: Novo Nordisk A/S, Jesper Westly). The other is entitled *Translating NovoSeven®️*. Speciale i Informationsvidenskab af A. Lihn Jørgensen og N.R. Thykier Videbæk. Aarhus University April 2007.

12.1 *THE NOVOSEVEN®️ STORY* (NOVO NORDISK 2004)

In this book, the authors give a chronological overview of the project's history from 1983, when Novo Nordisk decided to develop and promote two potential products for the treatment of thrombosis, namely low molecular weight heparin (LMWH) and tissue plasminogen activator (tPA). These plans led to my recruitment in 1983, with the task of establishing a research laboratory within the hemostasis area. The book describes how the Chief Executive Officer (CEO), Mads Øvlisen, and the Director of Research, Ulrik Lassen, enthusiastically listened to my ideas about FVIIa. Resources for the purification of a limited amount of

Treating Life-Threatening Bleedings.
DOI: http://dx.doi.org/10.1016/B978-0-12-812439-0.00012-5

plasma-derived FVIIa were granted. These were used to confirm the positive results I had published earlier [1] and led to the approval of a project, started in 1985, with the aim of developing a recombinant FVIIa for treatment for hemophilia patients with inhibitors (Fig. 12.1).

The biotechnical challenge is presented and how, in 1985, it was not known whether recombinant technology could be used for the production of such a complicated product as FVII. This section of the book also includes a description of Professor Earl Davie and his research group at the University of Washington in Seattle, USA, who had cloned and sequenced several human coagulation factors. Earl Davie was one of the founders of the biotechnical company ZymoGenetics. On June 1, 1984, Novo Nordisk signed a contract with ZymoGenetics for the development of a technique to produce recombinant FVII. In the Novo Nordisk book, it is pointed out that ZymoGenetics at the time already had considerable experience of mammalian cells and recombinant technology from their work with Novo Nordisk on the rtPA project.

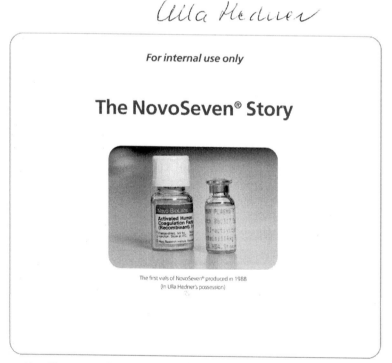

Figure 12.1 The cover of *The NovoSeven[R] Story*, published by Novo Nordisk, 2004.

The complete DNA sequence of human FVII was determined at Novo Nordisk by a research group in the "rDNA factor VII" project that started in January 1985. In the initial work, a cell line producing a hybrid molecule between FIX and FVII was used. The complete DNA sequence for human FVII was sent from Novo Nordisk to ZymoGenetics around Christmas 1985, and the final cell line for the production of human FVII was delivered from ZymoGenetics to the Novo Research Institute in January 1986. The continued development of the purification and characterization of the recombinant FVII molecule is described and in March 1987, 2 months ahead of time, a pilot production plant for rFVIIa was ready.

The continued scaling-up process from a capacity of 20 L to the planned 100 L of the pilot plant appeared to be unexpectedly problem-free and in March 1987, batches of 2000 L were being produced. A completely new rFVIIa production plant was built in Kalundborg and the production of rFVIIa started there in 1990.

The further story describes how the production was completely under control by the end of 1993 and how the preclinical testing started. However, in a note it is mentioned that after the merger between Novo Industri and Nordisk Gentofte, the production of rFVIIa was given lower priority in favor of the production of growth hormone. Also mentioned is the fact that most of the activities around the production of rFVIIa, such as testing for quality and stability as well as preclinical safety and efficacy were moved from Kalundborg to Gentofte. It also points out that this unavoidably led to difficulties in the allocation of resources between rFVIIa and the growth hormone.

Furthermore, *The Story of NovoSeven*® describes the clinical development of rFVIIa, which started in 1988. Part of this was the extended organization that was developed for the rapid delivery of rFVIIa to patients treated within the so-called "Compassionate Use Program." Several authentic patient stories from the "Compassionate Use Program" are also referred to. In addition "The NovoSeven® Expansion Programme" is described, illustrated with a description of the first patient from Israel who was given NovoSeven in connection with a high-velocity rifle injury. The patient developed a heavy bleeding and despite massive transfusion, his death appeared inevitable. After one dose of NovoSeven the bleeding slowed down allowing work on surgical hemostasis and the patient survived.

The product name NovoSeven was introduced in 1993, and the collection of data for license application was started. This finally added up to 63 files of documents, which was twice as many as for any previous Novo Nordisk product. The application was submitted in 1994 and, in 1996, NovoSeven was approved in Europe.

Finally, the mechanism of action of NovoSeven is described, underlining the fact that the entire coagulation system was partly rewritten as a result of the new findings made, due to the access of rFVIIa protein. This enabled scientists at Novo Nordisk in Copenhagen to collaborate with international research groups on elucidating the role of FVII in the hemostasis process.

12.2 *TRANSLATING NOVOSEVEN*®. SPECIALE I INFORMATIONSVIDENSKAB AF A. LIHN JØRGENSEN OG N.R. THYKIER VIDEBÆK. AARHUS UNIVERSITY APRIL 2007 [2].

This examination paper, written in Danish, starts by describing the development of rFVIIa as a successful innovation defined as "the effort to develop a previously known element for practical—commercial use and have it accepted, i.e., a form for the conversion of invention from idea to reality." The authors have used a translation model by Bruno Latour based on the idea that scientific facts are not "subjective (mental) representations of objective (natural) conditions but on the contrary are chains of the representations. These representations are called translations."

According to Latour "the basic motive in the understanding of an innovation based on translation, thus, is: no movement without change." It is also emphasized that "the good idea does not of itself flow through mind and matter." According to the translation model, the original idea can pass through a chain of changes that make the final result substantially different from the original idea.

In their analysis of the development of rFVIIa, the authors focus on a number of events that appeared important for the development of NovoSeven in the light of the translation model. They emphasize that, with this model, it is possible to avoid reducing earlier activities on the basis of later acquired knowledge. Furthermore, they point out that this method illustrates the central role of the so-called "articulation work" in the development of an innovation project. The opposite of the translation model is the so-called "diffusion model," according to which, earlier activities in the development are reduced to achieve a uniform and convincing description of a project's history. According to the authors, *The NovoSeven*® *Story* (Novo Nordisk) is an example of a diffusion model.

In their analysis of the NovoSeven project, Linn Jørgensen and Thykier Videbæk noticed that, according to *The NovoSeven*® *Story*, the development of NovoSeven is described as starting with my recruitment to Novo Nordisk in 1983 and then continues with the successful use of pure plasma-derived FVIIa carried out by myself and Walter Kisiel at the hemophilia clinic in Malmö. The development up to 1996, when NovoSeven was licensed in Europe, is described as a straight line. NovoSeven had by this time earned its first billion DKK. According to Linn Jørgensen and Thykier Videbæk, it became obvious to the company that NovoSeven had a larger potential than was first imagined. The successful treatment of an Israeli soldier with severe injuries whose life had been saved by NovoSeven contributed, according to the authors, to changing the strategic development to "achieve clinical acceptance of NovoSeven as a general hemostatic drug."

The continued analysis by Lihn Jørgensen and Thykier Videbæk points out that the surprise of the FVIIa project members was mainly due to the decision to use NovoSeven in bleedings other than in hemophilia patients and not in the fact that it worked. In this connection, Lihn Jørgensen and Thykier Videbæk quote a statement I made, "Ulla Hedner says in an interview, that already when we started the project in 1985, we knew, or at least I was convinced, that it *(faktor VII, red.)* would highly likely be effective in other bleedings." This impression was confirmed in other interviews.

In fact, already at the beginning of the 1990s, plans had been made for clinical studies in other indications than hemophilia. Novo Nordisk already in 1985 had applied for and been granted a patent for the use of FVIIa in cases of hemorrhages caused by too few platelets or bleedings in patients with an impaired platelet function. Also severe gastrointestinal hemorrhages and heavy bleedings in connection with serious traumata were covered by the patent applied for in 1985. Lihn Joergensen and Thykier Videbæk point out "that NovoSeven in liver diseases often complicated by bleedings." Against this background, Lihn Jørgensen and Thykier Videbæk question the real reason behind an FVII project at Novo Nordisk in 1985. They find no answer in the *NovoSeven*® *Story* and start their own investigation that led them to apply the translation method on the *NovoSeven*® *Story*.

The starting point for the authors is the question of why Novo Nordisk at all considered to start a project to develop recombinant FVIIa when other pharmaceutical companies did not find it to be a sufficiently good idea to develop rFVIIa for the treatment for hemophilia patients with inhibitors. In the interviews carried out with people within Novo Nordisk who had been involved in the project or been part of the company management at the time, a number of various reasons emerged.

Lihn Jørgensen and Thykier Videbæk draw the conclusion that the FVIIa project was considered to have much in its favor. They find two main reasons for embarking on the project:

1. The big hemophilia market, although everybody interviewed seemed to agree that the prospect of a commercial success, hardly could be the reason for starting the project.
2. The desire to expand the hemostasis portfolio beyond LMWH and tPA and, by doing so, establish another commercial area in addition to the insulin production.

They note a number of individual reasons (e.g., new cell technology) to strengthen the company as a frontline organization regarding research and development but found it impossible to discover more about the real reason why the project started. They therefore chose to follow the recommendation of Latour, originator of the translation method for processes: "when the complexity becomes confused, follow the actors!"

Their continued analysis deals with how the rFVII project became a translation project. This occurs when a problem changes from one context to another.

As already mentioned, from its start in 1985, the rFVII project at Novo Nordisk was characterized by a number of different interests. According to Lihn Jørgensen and Thykier Videbæk, it can, at this stage, be described as a chain of translations forming a complicated network of interests around the establishment of an FVII project. The next step, which makes the project more real, is the development of a recombinant drug that is found to be effective in preventing hemorrhages in a hemophilia patient in 1988.

The analysis then points out how the FVII project changes around 1989 mainly due to three different events. The first of these was the treatment for the first hemophilia patient, which partly changed the surrounding network and, consequently, also the rFVIIa itself. The authors point out that the reason the treatment could be carried out before it was approved for clinical studies was my background as a hemophilia doctor in Malmö and, basically, had nothing to do with rFVIIa. The successful treatment in Stockholm was significant for how rFVIIa became regarded internally at Novo Nordisk. As a result of this, rFVIIa was clearly linked to the treatment for hemophilia patients with inhibitors.

According to Lihn Jørgensen and Thykier Videbæk, the second event, which was crucial in limiting the area of use of rFVIIa to hemophilia, was the application for Orphan Drug status in 1988. The authors stress that the advantage with the application was that Novo Nordisk saw a possibility to bypass the Baxter Travenol patent for the use of rFVIIa produced from plasma. During the discussions with the FDA, it became obvious that to be given Orphan Drug status, the area of use must be limited to a smaller group of patients, which, according to the documents Lihn Jørgensen and Thykier Videbæk had available, resulted in putting a potential use of rFVIIa in other bleeding situations on ice for the time being.

The third event that took place in 1989 and, according to Lihn Jørgensen and Thykier Videbæk, played a crucial role for the continued development of rFVIIa, was the reorganization of Novo Nordisk. As a result the company was divided into three separate units, each of which was to be self-financing. Resources to projects in progress were allocated based on the income generated by each division's business area. This led to the establishment of a functional structure instead of a project–oriented one. The analysis describes the consequences for the FVII project and for hemostasis in general. The satellite structure, with scientists from different areas who collaborated on the FVII project, was dissolved, and the individual scientists were relocated in different functional areas. According the Lihn Jørgensen and Thykier Videbæk, the consequence for the rFVIIa project was a period marked by conflicts, partly general problems with cooperation between two different organization cultures and partly different opinions about the allocation of resources for the various projects. A conflict between me and the rFVIIa group on the one side and Nordisk Gentofte with the divisional management on the other side was noted. The view as to whether the new organization strengthened or weakened the project obviously varied noticeably within the group interviewed by Lihn Jørgensen and Thykier Videbæk.

Furthermore, their analysis shows that rFVIIa during this period, changed into being a project approaching completion and thereby ready to contribute to the income of the division. The quick licensing of rFVIIa for the treatment of hemophilia with inhibitors was therefore most appealing. They point out that with the changed context, the network necessary for continued development also changed.

The next phase in the story of rFVIIa is described by Lihn Jørgensen and Thykier Videbæk as "the project under attack." In their analysis they point out that Novo Nordisk, from 1989 to 1995, mainly focused on coordinating the insulin activities. It had gradually become obvious that Novo Nordisk had too many projects for the available resources. A reorganization process was started with the help of the McKinsey consultant company. On the list of "core areas" that was drawn up, rFVIIa was nowhere to be found. Despite this, it was "the single hemostasis project allowed to survive," as Lihn Jørgensen and Thykier Videbæk expressed it. The LMWH project was sold to Løven (Leo Pharma), despite the fact that a successful large clinical study just had been completed. According to the interviews, several colleagues believed that it was probably the opinion of the upper management at Novo Nordisk, which had saved the rFVIIa project from being killed.

Among the various interests that, from the beginning, had supported the rFVIIa project, only "interest for the patient" was retained. It was pointed out that this aspect had become more visible because of the many patients successfully treated with rFVIIa during the early 1990s.

A new reason for keeping the rFVIIa project emerged in the 1990s, namely the construction of a production plant dedicated for use of gene technology and complex molecules. In Lihn Jørgensen's and Thykier Videbæk's analysis they stress that rFVIIa, with the production plant as a symbol, at this time, had "gained so much in strength that it was now the surroundings which were forced to adapt to factor VII."

The continued development of rFVIIa is characterized by the expansion of production and, according to Lihn Jørgensen and Thykier Videbæk's analysis, led to the involvement of several new actors whose job was to stabilize and standardize the production of rFVIIa.

According to Lihn Jørgensen and Thykier Videbæk, 1995 saw the end of the rFVIIa project that was started in 1985. When NovoSeven was licensed in 1996 for the treatment of hemophilia with inhibitors and hemostasis research eradicated in 1995, the possibilities for the further development of rFVIIa for use in a wider area of indication disappeared. At the same time the product NovoSeven was strengthened. The project rFVIIa had been transformed into the product NovoSeven, which was now an independent actor able to treat patients "in all of Europe, without the assistance of Novo Nordisk and not the least, it was able to make money."

In the translation analysis, it was suggested that "the real research on factor VII and the coagulation system at Novo Nordisk was finished" with the introduction of divisions in 1989. It is pointed out that this led me to intensify my relations with the scientific community outside Novo Nordisk and that the NovoSeven project basically survived on the basis of my personal scientific contacts "established before my recruitment to Novo Nordisk." An influential network was formed consisting of hemophilia doctors and scientists within the hemostasis area, who all contributed to the establishment of rFVIIa.

According to Lihn Jørgensen and Thykier Videbæk, this external network meant more for the project than just "a helping hand when in need." In their analysis, they stress the importance of the symposia arranged every second year with a start in 1987. These were held at Novo Nordisk in Copenhagen and provided an opportunity to exchange clinical experience and scientific results within the hemostasis area. The mutual exchange of experiences meant that the work at Novo Nordisk was not confined to Novo Nordisk. There was an ongoing dialog between the project group and hemophilia doctors outside Novo Nordisk.

As part of this collaboration I held a lecture with the title *NovoSeven*® *as a universal hemostatic agent* at the Symposium in May 1999. Two months later, the first patient was successfully treated in connection with a severe gunshot wound. It is pointed out by Lihn Jørgensen and Thykier Videbæk that "when an Israeli hemophilia doctor is confronted with a soldier who is bleeding to death as the result of a gunshot wound and chooses to use NovoSeven to stop the bleeding, it was not a decision taken out of the blue. What for some, came as a surprise, had been obvious for a long time for others."

The last chapter in Lihn Jørgensen and Thykier Videbæk's work has the title *Articulation*. It analyzes my role in the rFVIIa project as they cannot find an immediate explanation for my becoming, in Novo Nordisk's history, the principal person in the project. They start by pointing out that the rFVIIa project has "all the time been dependent on its ability to interest others," and they describe how the project has passed through a "long series of translations." They also point out that the fact that the idea of rFVIIa was mine cannot explain why I came to play the main role.

"The idea in the translation model is that one actor is like all the others and has therefore no primary status which can by itself serve as an explanation for the following process."

Several of the interviews stressed the following qualities in me, important for keeping the project together: (1) "an understanding of the patients who, in her opinion, were in need of a factor FVII-product," (2) "her gift as a communicator," (3) "her knowledge of the clinical treatment of hemophilia," and (4) "her role as a generalist. She showed understanding and interest in all parts of the process, without, being an expert in all areas herself." In the management interviews it was stressed that a condition for the success of the project was that "it had an advocate." The most important qualities emphasized in this connection were that "she saw a reason with her work. She was also an enthusiast. That was also important. And she was persistent. And it was not one year. It was not two years. It was many years. At the start it was like wandering in the wilderness. Persistence was necessary, extraordinary persistence."

Colleagues in the project group stressed the importance of "having someone who understood the clinical indication and the patient's situation." Clinical experience was important and "worldwide contacts" as well as one "who has the physiological and clinical understanding of the patient and the environment in which the product is to be used …." Another important factor underlined was the extensive flow of information maintained between the different specialist groups, which contributed to motivate colleagues. "If one gets quick feedback on what one is doing, if it is working or not working, and can see it as part of a whole, it is extremely encouraging." "It also introduced a sense of belonging. We all realized that it was not just a question of pushing a bucket of something through the door a couple of times a week. … but that there was an overall purpose that we could all identify with."

In summary, the research management emphasized that "there is only one person who is behind this. And it is Ulla. With compassion, mental strength and professional insight, everything is possible."

Lihn Jørgensen and Thykier Videbæk stress that there was a strength in my position and role at Novo Nordisk on account of my relation to the hemophilia clinic in Malmö and to the patients. Also it had been demonstrated that FVIIa did stop bleedings in hemophilia patients. They point out, that my role, by informing and by understanding the different problems and, above all, by describing the patients' situation, helped the various interests in the project to approach each other and by doing so made the project move forward toward the treatment of patients.

In the further analysis, the articulation work in the FVIIa project is described as being patient-centered. All those who were interviewed underlined that an important factor for moving the project forward was my understanding of the patients. Important for this process was (1) "effective communication between actors" and (2) "a common interest in the purpose and acceptance of the fact that the patients' suffering took priority over the individual actor's own agenda."

Of course, this work demanded numerous other activities that, in their analysis, Lihn Jørgensen and Thykier Videbæk call "hidden work" as they are never mentioned in the official NovoSeven® Story that concentrated on technical and laboratory problems.

In the final summary, it is concluded that the conclusive proof that NovoSeven can stop hemorrhages in other patients than those with hemophilia was the successful treatment of the Israeli patient with life-threatening bleedings caused by gunshot wounds. Based on the facts put forward in my lecture, NovoSeven® as a universal hemostatic agent, a dose of NovoSeven was administered. In Lihn Jørgensens and Thykier Videbæk's analysis, this event is described as a surprise for Novo Nordisk although not to me or my colleagues in the project. They point out that this treatment was the ultimate proof that I was right in my conviction that NovoSeven could be used as a universal hemostatic agent,

Furthermore, it was concluded that the activities which the project group considered crucial for its success and which are summarized in the concept of articulation work could never be considered equally important as other activities. This led to my role in the official description being limited to being the originator of the idea of rFVIIa. The other work important to the project remains concealed for all, except those who were involved in the development process. The translation analysis asks why, in my description of the development of rFVIIa in The NovoSeven® Story, I have left out the work that made me the central figure in the project. Their explanation is that the project's history has taken the form of a diffusion story, presenting a rational work proceeding from hypothesis to production. There is no room for the necessary articulation work. The price for creating this clear-cut story, according to the authors, was a high one for me to pay, as the entire work of "giving meaning to the project—the articulation work—is missing. The same applies to her credit for the treatment of the wounded Israeli soldier in 1999, and not least for the fact that, already in 1986, she saw what would make NovoSeven® such a great innovation fifteen years later." They point out that it is my own description and stress "that she was willing to pay this price underlines exactly why she has the leading role in the story of NovoSeven®. For more than twenty years, she has been willing to make sacrifices in order to treat patients who had no other options."

Lihn Jørgensen and Thykier Videbæk's analysis ends in the year 2000, when Novo Nordisk started their expansion program for NovoSeven as a general hemostatic agent. By this time, NovoSeven had just earned its first billion DKK.

12.3 A DRUG FOR TREATMENT OF LIFE-THREATENING BLEEDINGS: THE DEVELOPMENT OF rFVIIa—BY ULLA HEDNER

The third description of Novoseven's history is my own, which describes the process as I perceived it. My presentation is based on various sources and on my own experience.

1. Written material from my time at the Clinic of Hemostasis disorders in Malmö, research results, documents from the preparation of FVIIa from human plasma 1980−81.
2. Written material consisting of various summaries, investigations, decisions, memos, copies of registration material from the different developmental phases of rFVII within Novo Nordisk.
3. Publications including research results and reviews of relevant literature.
4. Treatment of patients, including surgical history.
5. My own recollections partly written down.
6. *The NovoSeven® Story* (2004) and *Translating NovoSeven®*. Speciale i Informationsvidenskab af A. Lihn Jørgensen og N.R. Thykier Videbæk. Aarhus University April 2007.

Thus, my presentation is based on various, critically examined, sources and my own participation in the process. Consequently, it is dependent on my personal perspective (Fig. 12.2).

Figure 12.2 Ulla Hedner in her office at Novo Nordisk, 1994.

12.4 A COMPARISON OF THE THREE DESCRIPTIONS

The descriptions have a great deal in common. I recognize myself in the first two. I have also contributed to the first description and participated in the interviews of the second one.

The NovoSeven® *Story* gives a detailed description of the technical aspects of the developmental work whereas *Translating Novoseven*® describes the background from Novo Nordisk's point of view as to why they at all embarked on the project of developing rFVIIa. It then focuses on how an innovation is converted from an idea to reality. This description puts great emphasis on the actors and the development of the actual process. It is more detailed than the first description.

In many respects, my description is closest to the second one, *Translating NovoSeven®*, which, as well as on the actual process, focuses on the actors. My description also gives place to those who have taken part in the development. My presentation confirms several of the observations made by the authors of the translation analysis. With the help of their presentation, I have also seen my role in the development of NovoSeven more clearly.

Translating NovoSeven® places great emphasis on the articulation work and its importance in the development of NovoSeven. This had not occurred to me as I have seen the process from my point of view, whereas *Translating NovoSeven®* describes it from the viewpoint of colleagues and management, which gives it another perspective. I have no great objections to their descriptions, which I recognize. A few things stand out more in my presentation, e.g., my description follows the development of NovoSeven further on in time, covering the years up to my retirement in 2009. *Translating NovoSeven®* ends around the year 2000, with the establishment of the NovoSeven Expansion Program.

My presentation also differs from the description in *Translating NovoSeven®*. It focuses more on the actors and change, whereas mine devotes more space to the problems encountered during the process, which is only natural from my perspective.

REFERENCES

[1] Hedner U, Kisiel W. Use of human factor VIIa in the treatment of two haemophilia A patients with high-titer inhibitors. J Clin Invest 1983;71:1836—41.
[2] Lihn Jørgensen A, Thykier Videbæk NR. *Translating NovoSeven®*. Bruno Latour's model of translation as a framework for analyzing the innovation project NovoSeven®. Speciale i Informationsvidenskab, Institut for Informations og Medievedenskab, Humanistisk Fakultet, Aarhus Universitet, Denmark. April 2007. Quotations of the Danish text are translated into English by the author.

CHAPTER THIRTEEN

Some Final Remarks

Contents

Before summarizing some of the conclusions and lessons learned from the development of rFVIIa for treatment of life-threatening bleedings, I would like to describe the environment from which I come.

The Professor of Internal Medicine, Jan Waldenström, was recruited in the early 1950s to the General Hospital of Malmö, part of the medical faculty of the University of Lund, Sweden. Among other things, he taught his medical students to pay careful attention to their patients, from whom they would be able to learn directly how to approach further research to elucidate their diseases. He stressed the importance of the listening and conversation with the patient for diagnosing and therapy. His favorite expression was that patients show you the shortcuts to medical solutions of unknown symptoms and diseases. The interest in each human being was combined with intellectual curiosity that made him see a biological problem in each patient, which he saw as a challenge to solve. He himself was the discoverer of several hematological diseases such as Waldenström's macroglobulinemia, inherited porphyria, the carcinoid syndrome, and autoimmune hepatitis.

From Jan Waldenström I, thus, learned to focus on the patient. He also stressed the importance of collaborating with biochemically educated scientists in the solution of clinical problems, which gave me another of my basic ideas, the necessity of close teamwork between clinicians and biochemists. In this context I will especially mention the excellent description of Jan Waldenström given by one of his students from 1970, Ingemar Turesson [1]. The combination of Waldenström's great interest

Treating Life-Threatening Bleedings.
DOI: http://dx.doi.org/10.1016/B978-0-12-812439-0.00013-7

in human beings and an intellectual curiosity led him to identify biological problems as challenges requiring solutions. Along these lines he
inspired many of his students. According to Ingemar Turesson his motto
was "from bedside to benchside."

Another important part of my background is the Department of
Bleeding Disorders started by Professor Inga Marie Nilsson in the middle
of the 1950s. She had spent some time in the laboratory of Erik Jorpes,
specialist in medical chemistry at Karolinska Institutet, Stockholm. In his
laboratory, Margareta and Birger Blombäck worked on the purification of
coagulation proteins. The Blombäcks identified one protein fraction in
which the FVIII protein missing in hemophilia A patients was concentrated and thus present in higher amounts than in the original human
plasma. Inga Marie Nilsson had a position at the Clinic of Internal
Medicine of the General Hospital of Malmö where she took care of
hemophilia patients. She therefore was able to test the concentrated FVIII
fraction prepared at Karolinska Institutet, Stockholm, in patients. The
collaboration between her and the Blombäcks at Erik Jorpes department
led to the start of a treatment center for hemophilia patients at the
Malmö General Hospital in the middle of the 1950s, as a small part of the
Clinic of Internal Medicine. This later became one of the world's most
advanced hemophilia clinics (Fig. 13.1).

During my years in Medical School I spent time in this environment
during my vacations working as a laboratory technician and later as a
nonpaid scientist. This contributed to a basic knowledge of the bleeding
disorders and their treatment, as well as the techniques necessary for diagnosing them. It certainly made me feel fully convinced about the importance of combining clinical work with scientific knowledge. I want to
summarize my experiences in three important areas.

13.1 PATIENTS AND CLINICAL FOCUS

The patient and clinical focus were from the beginning important
for me on several accounts and have remained so during the whole development of rFVIIa.

My close interest in the patients generated a strong feeling with regard
to the need for an improved treatment for the hemophilia patients who
had developed inhibitors against the blood protein they lacked. For me it

Figure 13.1 Professor Jan Waldenström (sitting) with two of his younger colleagues at Malmö University Hospital, Inga Marie Nilsson and Carl-Bertil Laurell. *Photograph by Rune Ohlsson in 1995* (Sydsvenska Medicinhistoriska Sällskapet, Björn Henrikssons samling).

was a shame that we had nothing better to offer these patients than the activated prothrombin complex concentrate available already at the time.

The same interest kept me continuing despite all the obstacles that arose during different periods of the development of rFVIIa. Thus, during the difficult 1990s, when the whole existence of rFVIIa at Novo Nordisk was questioned, it was the patients that kept me going. At that time I had seen several patients' lives saved with the help of rFVIIa. I, thus, was convinced that there was a need for the drug in the clinic. In the dark moments it was a clear help to be reminded of all the grateful patients around.

I also want to stress the importance of the patients and the clinical work in the process of understanding the mechanism of action of rFVIIa and of the entire tissue factor (TF)—dependent hemostasis. This goes for both the experimental work and the clinical effect.

An example of the importance for me to relate the experimental models to the clinical treatment of hemophilia is the development of the cell-based model in collaboration with the Chapel Hill research group. The formation of low and slow thrombin generation in the absence of the hemophilia proteins observed in the model with TF-expressing cells and platelets seemed to me relevant to the clinical picture in hemophilia patients. Furthermore, the initially limited thrombin generation in the presence of TF-expressing cells only resulted in an impaired thrombin burst as demonstrated in the model. This fitted well with the clinical picture in hemophilia patients who form loose and fragile clots, easily and prematurely dissolved by protein degrading enzymes around a site of injury. In the presence of a defective thrombin generation, such loose and fragile clots are formed, which were clearly illustrated in our model.

From the clinical picture I can define three patients, who I would like to call "my three key patients." They have each of them contributed substantially to promote the knowledge of the mechanism of action of rFVIIa and also contributed to increased knowledge of the whole hemostasis process.

Patient 1 was the boy who received the plasma-derived FVIIa when losing a primary molar. He was the one who convinced me that the idea to use extra FVIIa to stop bleedings in hemophilia was relevant, "the proof of principle."

Patient 2 was the one who underwent orthopedic surgery under cover of rFVIIa. He was the first patient receiving rFVIIa. The treatment was performed with a license for use in this special patient with an urgent need for surgery and lack of effective treatment. No bleeding occurred, which gave me the "proof of concept," because I knew from my own experience that hemophilia patients cannot undergo orthopedic surgery without bleeding problems unless a hemostatic drug is used.

The experience in patient 2 led to the use of rFVIIa in several hemophilia patients with inhibitors and "life- and limb-threatening bleedings," according to the "Compassionate Use Program" as advised by FDA. The successful use of rFVIIa in these patients convinced me about the hemostatic effect of rFVIIa and of the clinical need for such treatment. This helped me to stay on and continue working on piloting rFVIIa through the complicated and important development toward registration in the 1990s.

Patient 3 was the boy whom I met when I was back in the clinic in the late 1990s. He had been using rFVIIa for some time and was judged to be "a low-responder" to rFVIIa. He had developed contractures in both his knee joints confining him to a wheelchair despite having been treated with rFVIIa.

Being back in the clinic gave me the opportunity to try NovoSeven, registered since 1996, in patients. Together with the other Swedish hemophilia doctors, I started by performing a preliminary study of the plasma half-life of rFVIIa in children in the six inhibitor patients available in Sweden. The boy with the knee contractures had a clearance rate of rFVIIa from plasma three times as rapid as in adults. It seemed obvious to me that he may need a higher dose for efficient effect of rFVIIa. Together with my colleagues at the hemophilia clinic in Malmö, I introduced a doubling of the dose he had received, which turned him from a "low responder" to a "high responder" to rFVIIa. This supported the signs I had observed that a higher dose than initially thought, was required, at least in some patients, to reach a satisfactory effect. The use of a higher dose of NovoSeven and later the studies of a prolonged effect of rFVIIa thus were results of observations in this patient. This boy also underwent orthopedic traction treatment resulting in straight legs. Thus, he was able to leave the wheelchair.

The observations in this patient helped to move science forward regarding the distribution of injected rFVIIa. Studies in mice have so far demonstrated localization of injected rFVIIa to various tissues. In some of them the rFVIIa could be traced for up to 7 days after the injection. Also, the FVIIa-TF complexes were demonstrated to be active. These findings suggest that the use of rFVIIa may prevent bleedings in a long-term setting and thus would be effective as prophylaxis in hemophilia patients with inhibitors. Actually, smaller clinical studies in several countries have supported the effect of regular injections of rFVIIa three times a week in preventing bleedings.

13.2 DEVELOPMENT OF A HEMOSTASIS PRODUCT IN AN INSULIN COMPANY

During the work on purification of FVIIa from human plasma, it became obvious to us that to be able to provide a product for use in a larger population of patients in need for an improved therapy, the competence of a pharmaceutical company was necessary. After having been recruited by Novo Industry A/S to establish a research laboratory within hemostasis, it became clear to me that Novo Industry had all the knowledge and techniques required. Based on various reasons the company decided to start a project to develop recombinant FVIIa for use in hemophilia patients with inhibitors against the proteins they are lacking. It was approved in June 1986.

The satellite model used by Novo Industry in research projects turned out to be excellent. A small group of scientists with various specialties was put together. This helped to create focus on the goal of developing a product to help patients with hemophilia complicated by inhibitors to a better life. My role was to link the group to the clinical reality of these patients. The atmosphere within the group was characterized by openness, energy, and dedication. Without the competencies gathered in this group the development of rFVIIa would never have been possible.

Another very important part was the special competence within the fermentation of living organisms present at the Novo Industry A/S where industrial enzymes had been produced for many years. This knowledge was a necessary requisite for the successful and quick development of the technique for making baby hamster kidney cells produce a complicated human protein like FVIIa. Also the personal dedication and engagement in this work added strongly to the success in having rFVIIa ready for use in patients within 3 years.

The importance of marketing a new product, especially one in a new therapeutic area, has become obvious to me. Marketing does not occur by itself. It is clear that knowledge about the clinical aspects is important for marketing a product. It is also obvious that such knowledge in a company having experience in diabetes for many years cannot be as extensive in hemostasis. This actually became clear to me very early in the development of rFVIIa. Already from the beginning I had to spend a lot of energy and time on what I used to call "internal marketing." The future "external marketing" seemed easy to me because I knew the need for an improved therapy in hemophilia patients with inhibitors.

In the "internal marketing" I had a lot of help from my international network including scientists interested in hemostasis and especially in the TF-dependent part of it, as well as clinicians with a need for improved treatment of their patients.

The international support consisted of individual research collaborations. In an attempt to maximize the effect on the internal level of knowledge I initiated the symposia with 1 day devoted to research and the other day to clinical aspects. I thought this to be a way to spread information and new knowledge widely to all of the Novo Nordisk people involved with hemostasis and the development of rFVIIa. To the symposia, I invited international scientists and clinicians active in the field. This also stimulated the establishment of individual research collaborations that turned out to contribute significantly toward making our own research group as one of the most competent in the area. A bridging role in facilitating the understanding of the mechanism of action of rFVIIa among external scientists was played by Pim Tiljberg of the Novo Nordisk affiliate in the Netherlands. His talk about "scientific marketing" helped substantially.

The knowledge about the importance of marketing and its need for people believing in the product they are going to sell is the background of my increased engagement in Novo Nordisk affiliates after NovoSeven was approved. I realized that if the launch of NovoSeven was to be successful, the affiliates needed help to learn about rFVIIa and not least about the disease and the patients with hemophilia. In this work certain individuals played an important role. It is clear that the success of NovoSeven has been dependent on leaders in different affiliates and business areas who have been able to change the attitude toward a hemostasis product in the Novo Nordisk portfolio. I have mentioned the important role of Jesper Højland in the marketing of NovoSeven in France in the late 1990s. Important roles in this context have also been played by Eric Wong who turned the Taiwanese hemophilia doctors into some of the most dedicated users of NovoSeven. Another area that has turned into one of the more enthusiastic ones with regard to hemophilia treatment is the Gulf area including countries such as Saudi Arabia, Kuwait, United Arabic Emirates, and Oman. This has been made possible due to an excellent Novo Nordisk team and not the least to the support from Mads Bo Larsen (Fig. 13.2).

In the above I have tried to illustrate the importance of close teamwork among people of the company and external parties. The latter are not least important when it comes to launching and marketing the product. In fact, this external support from clinicians helped to push the development of rFVIIa forward. The need for the product in clinical work was obvious. Much work has to be continued long after the product has been approved and this requires attention for an extended period of time. This phase of the development is threatened by changes of the people who were responsible, reorganizations, and internal conflicts. It should be emphasized that even the best product may fail in this process!

During the whole development process there is a need for a "driver." The need for a "cognitive motor" was stressed by Ivan Östlund in his

Figure 13.2 A meeting on rFVIIa at Dubai, UAE, November 2015. Novo Nordisk employees and the author in the middle.

work on the development of Losec (omeprazole), which was closed five times during its development due to the difficulties for the business part of the company to see the advantages of a new product in a new therapeutic area [2]. In the examination paper by Lihn Jørgensen and Thykier Videbæk entitled *Translating NovoSeven*®, they stress the importance of what they call "the articulation work" carried on by me during the whole development. The details are given in Chapter 12, Three Different Descriptions of How rFVIIa Was Developed. The important part according to the authors was my relation to the hemophilia clinic and to the patients. This made my articulation work "patient-focused" and helped to keep "a common interest in the purpose and acceptance of the fact that the patients' suffering took priority over the individual actor's own agenda." My analysis of the relevant documents confirms the importance of the articulation work.

13.3 OBSTACLES AND PROBLEMS

It is generally well known that reorganizations and fusions between pharmaceutical companies create problems and usually delay the development of new drugs. What were the experiences of the influence of the fusion between Novo Industri A/S and Nordisk Gentofte on the development of rFVIIa? This fusion occurred in 1989 at the time the clinical development of rFVIIa was about to start. The fusion was followed by an extensive reorganization dividing the company in different divisions meant to be self-supporting. The rFVIIa project was included in the Biopharmaceutical Division together with growth hormone and low molecular weight heparin. This meant the breakup of the satellite model used for rFVIIa, which had proved to be so successful. The different individuals of the group were included in the various functional units including pharmacology, toxicology, production, and assay department. The major project of the Biopharmaceutical Division, the growth hormone, became the dominating product mainly because it was already on the market and had a high status in the division mainly consisting of people from the old Nordisk Gentofte. In fact, the trust in rFVIIa as a future product of any value was weak.

As a result of these changes the knowledge within the area of hemostasis was diluted. Many new people became involved in the planning of the future development of rFVIIa, everybody bringing in new ideas not necessarily useful in the delicate planning of the clinical development, which included the incorporation and follow-up of the already running "Compassionate Use Program."

My own role became restricted to research covering the whole biopharmaceutical area addressing new products. The rFVIIa project was considered a pure development project and regarded as beyond any need for research activities. However, the use of rFVIIa for stopping bleeding in patients with hemophilia and inhibitors against the factors these patients were lacking was a completely new concept of treatment. It was therefore clear to me that we still needed to work on the dosage and on the detailed mechanism of action.

The understanding of these aspects was not widely accepted in the new environment and I had a very little influence on the general opinion in the issues. In fact, I had no official authority to influence the further development of rFVIIa. To find resources for necessary work on the mechanism of action of rFVIIa, I had to initiate collaborations with external research groups such as the Chapel Hill Group.

The problems and delays occurring during the development of rFVIIa up to the final approval in 1996 reflect the difficulties in changing procedures during the developmental phase including new directions and attitudes to, among other things, the approval procedure.

It is not clear whether the time for approval (1990—96) would have been shorter if some of us experienced with the early development and background of rFVIIa as a hemostatic agent had been allowed to take part in the development of documents and handling of the approval process especially with regard to FDA. However, it is easy to see many obstacles that cropped up during this process.

My work with rFVIIa has taught me a lot: Clinic and experimental work has to be kept together by constructive collaboration, competent people, important articulation work, fundamental and maximal use of international research contacts and conferences, a competent and well-functioning pharmaceutical company, care taken with fusions and reorganizations, and necessary follow-up work after approval.

13.4 GENERAL DEVELOPMENT OF NEW DRUGS IN THE FUTURE

From my story of the development of rFVIIa, some conclusions could be drawn of potential use in general development of new drugs. In this context I would like to stress the following factors that I found to be important:

- Collaboration between scientists also outside the company. Not the least important is to include doctors working in the clinic, close to patients. These are the most apt to identify clinical problems in need for a solution.
- Close collaboration within the company between people with various competencies. A continuous sharing of information between production, pharmacology, clinical development, research, and marketing is necessary. Actually, the project goes back and forth between the different compartments because of various problems needing solutions.
- During the process a lot of fantasy and willingness to try various solutions is necessary, which requires openness between the different departments and individuals involved.
- Any developmental process includes a number of failures, which requires a good deal of persistence. In this context the need for "drivers" in a project becomes obvious.

During recent decades the question of whether the formation of global mega companies facilitates the creativity and successes has been more and more relevant. In this discussion it is often stressed that the background for the successful accomplishments of some of these global companies seldom is dependent on strong leadership or brave reorganizations, but rather the environment of research and technical progress and the presence of devoted individuals. In a review of a book describing the development of Pharmacia & Upjohn, the danger for a global company to end up in a rhetorically guided economy of plans, expectations, and estimations is stressed. Such development may be far away from the real-world economy dependent on innovations and products needed by patients [3].

REFERENCES

[1] Turesson I. Jan Waldenström—en föregångsgestalt inom Svensk hematologi. In: Westin J, Gahrton G, Hast R, Simonsson B, Turesson I, editors. Femtio år med svensk hematologi (Fifty years with Swedish Hematology). Göteborg: Svensk förening för Hematologi; 2011.
[2] Östlund I. "Magsårsmediciner" (Drugs for treating gastric ulcer). Från örtakok till läkemedel. Framgångar och bakslag I medicinernas värld under 50 år (From herbs to medicines. Successes and failures in the world of medicines during 50 years). Kristiandstad: Apotekarsocietetens förlag; 1991. p. 209—60.
[3] Benner M. Review of P. Frankelius, Pharmacia & Upjohn: Erfarenheter från ett världsföretags utveckling (Pharmacia & Upjohn: Experiences from the development of a global company), (Liber 1999), Svenska Dagbladet; 1999.

INDEX

Note: Page numbers followed by "*f*" refer to figures.

Printed in the United States
By Bookmasters